The Methane Security Problem

The Methane Security Problem

Climate Feedbacks as a Strategic Threat

By

Imani Z. Okere

Vij Books

New Delhi (India)

Published by

Vij Books
(Publishers, Distributors & Importers)
4836/24, 3rd Floor, Ansari Road
Delhi – 110 002
Phone: 91-11-43596460
Mobile: 98110 94883
e-mail: contact@vijpublishing.com
www.vijbooks.in

ISBN: 978-93-7794-008-9 (HB)

ISBN: 978-93-7794-009-6 (PB)

ISBN: 978-93-7794-010-2 (eBook)

This book was developed using advanced writing tools to support the drafting and ideation process. All content has been thoughtfully curated, refined, and reviewed to ensure quality and relevance.

Contents

Introduction

On 29 May 2020, a fuel tank collapsed at a power plant near Norilsk, in the Russian Arctic, spilling more than 20,000 tonnes of diesel into rivers and tundra, according to Russian authorities and reporting at the time. The ground beneath the tank had shifted. There was the metallic stink of fuel, the black-red sheen on water, the ugly soundless fact of containment arriving too late. A local industrial failure became a national event because it exposed something larger than a defective structure: the weakening of assumptions that had once sat quietly beneath the state's logistical confidence. A hairline crack had become visible.

The modern security imagination was built for countable things. Divisions could be tallied. Fuel could be rationed. Reserve stocks could be measured in days, not moods. Even uncertainty often came in familiar forms: an adversary's intent, a commodity shock, a harvest failure, an election upset. Methane feedbacks disturb that inheritance because they do not present themselves as a single enemy, a single event, or a single timetable. They alter rates. They shorten margins. They make background conditions less reliable, and they do so in ways that are difficult to attribute cleanly while the damage is unfolding. Timing changes everything.

This matters well beyond atmospheric science. When methane-related warming intensifies near-term pressure on ice, permafrost, heat, fire conditions, or hydrological extremes, it does not merely add one more environmental concern to an already crowded policy agenda. It reaches into the state's practical machinery. It changes how a runway settles, how long a winter road remains usable, how often a port can count on access, how emergency services calculate simultaneous strain, how insurers reprice exposure, how treasuries distinguish temporary repair from permanent redesign. What looks at first like an emissions problem becomes, at closer range, a problem of territory, infrastructure, procurement, legitimacy, and time.

That is why the language of externality is no longer sufficient. An externality can be acknowledged and priced at the margins. A strategic threat to state capacity is different. It bends the terms under which public authority performs its ordinary functions. It makes some places costlier to serve and some assets harder to trust. It forces governments to act under deep uncertainty while still being judged by electorates, markets, courts, and allies as though uncertainty were merely an excuse. The demand for precision does not disappear when confidence falls. It often becomes harsher.

There is a temptation, when speaking about climate risk in security terms, to reach too quickly for grand rhetoric: existential danger, geopolitical rupture, planetary emergency. Such language can be morally understandable and analytically lazy. The harder task is to trace how disruption actually travels. A thawing substrate deforms a storage site or road. That deformation changes inspection schedules, insurance terms, and repair windows. Delays raise costs. Costs alter budgets. Budget stress leads to deferred maintenance elsewhere. Service reliability falls. Residents lose confidence. Investors demand higher returns or state guarantees. A physical shift becomes an administrative one, then a fiscal one, then a political one. The chain is not theatrical. It is institutional. That is precisely why it deserves strategic attention.

Consider again the Arctic, where the phrase "thaw" can sound deceptively gentle. Permafrost is not a poetic abstraction. It is ground used to anchor buildings, roads, pipelines, depots, radar stations, storage tanks, and homes. When that ground becomes less predictable, maintenance ceases to be routine. It becomes a race against changing conditions and shrinking seasonal windows. In northern communities, sovereignty is exercised not through slogans but through services: medical evacuation, air access, fuel delivery, policing, communications, school operation, dependable mail. If a runway buckles or a road softens, the constitutional form of the state may remain intact while its felt presence weakens. People notice missed flights. They notice shelves. They notice silence.

The issue is not confined to remote regions or to dramatic failures. In August 2023, authorities in Canada's Northwest Territories ordered the evacuation of Yellowknife as wildfires approached. Nearly the entire city was affected. The immediate cause was fire, not methane. That distinction matters. So does the wider context: a warming world loads more stress into the systems that governments rely on to move people, issue warnings, sustain communications, and maintain order across long distances under concurrent strain. Smoke, road congestion, fuel supply, health services, accommodation capacity, aircraft availability, public messaging—each part leans on the others. Climate-related security risk rarely arrives in a single file. It compounds. A plan written for one hazard can fail because three other systems are already tired.

Methane sharpens this problem because it is bound up with near-term warming in a way that presses on decision windows. Carbon dioxide remains central to long-run climate stability. That is not in dispute. But security planning lives not only in end-of-century charts; it lives in asset lives, fiscal years, maintenance cycles, electoral mandates, insurance renewals, and the time it takes to procure steel, train crews, or redesign a port. Methane matters here because timing matters. A faster push on near-term warming alters the background against which long-lived decisions are made. The state does not buy infrastructure one season at a time. It makes promises into futures it cannot fully see.

This is where climate feedbacks create a particular strategic difficulty. A direct emissions source can be targeted, regulated, monitored, and debated within familiar policy frameworks, however imperfectly. Feedbacks are slipperier. They imply processes that may accelerate change beyond what linear assumptions would suggest, yet they do not grant policymakers the luxury of claiming certainty about thresholds, dates, or magnitudes. The language itself invites misuse. Non-linearity is often treated as permission for alarmism by one side and dismissal by the other. Neither response helps. The practical question is narrower and more demanding: how should institutions plan when the tails of risk matter, the causal picture is incomplete in real time, and public accountability punishes both false reassurance and visible overreaction?

The answer begins with a change in emphasis. Hazard is not the same as threat, and threat is not the same as vulnerability. A storm can remain a meteorological event until it collides with exposed housing, brittle grids, empty contingency funds, fragmented jurisdictions, or mistrusted warnings. The same applies to methane-related warming and its feedback effects. The atmosphere does not need hostile intent to produce strategic harm. Political intent is unnecessary when physical instability and administrative fragility meet. What turns a hazard into a threat to state capacity is the condition of the institutions asked to absorb it.

That point can seem almost banal until one looks closely at how modern systems have been optimised. Efficiency reforms, lean procurement, consolidated supply chains, just-in-time maintenance, tightly specified staffing, and narrow budget horizons can all appear prudent in stable conditions. They can also remove the slack that crises require. A resilience gap rarely announces itself with a single dramatic memo. It appears as a missing spare part, a vacant technical post, an ageing bridge deck left one more year, an emergency contract that cannot be activated because the authorised signatory is unreachable, a mutual aid agreement that works on paper until neighbouring jurisdictions are hit at once. The state does not fail all at once. It frays at the seams.

Nowhere is this fraying more revealing than in the politics of information. Methane is notoriously difficult to fold into neat narratives because baselines are contested, sources and sinks vary, monitoring technologies have limits, and inventories, atmospheric measurements, and models do not always align in ways that calm public debate. That does not mean nothing is known. It means that "how much" is often inseparable from "how known". Satellites can widen visibility. Sensors can sharpen local detection. Scientific cooperation can improve attribution. And yet information without clear decision rights often breeds frustration rather than competence. A ministry receives warning but lacks the mandate to act. A municipality sees exposure but cannot fund redesign. A community reports changes that do not fit the reporting template. Data accumulates; authority stalls.

Still, caution is necessary. Not every damaged road in the north, not every fire season, not every supply failure should be folded into a methane security narrative. Some infrastructure is old. Some governance is weak for reasons unrelated to climate. Some engineering performs remarkably well under difficult conditions. The Trans-Alaska Pipeline System, for example, was built with elevated segments and design features intended to manage unstable or sensitive terrain, and decades of maintenance and adaptation have kept it operating through conditions that would defeat cruder infrastructure. This is not a reason for complacency. It is a reason for discipline. The strategic claim is not that feedback risk renders planning impossible or that every failure is climatically driven. It is institutions that acknowledge uncertainty, build adaptive pathways, and invest in maintenance, redundancy, and local knowledge that will absorb shocks better than those that treat past stability as a continuing entitlement.

That distinction matters because the security field has its own habits of distortion. The phrase "threat multiplier" has often been used to draw attention to climate risk, and sometimes usefully so. Yet it can also blur more than it clarifies. Multiplying what, exactly? Through which institutions? At what point does a drought, thaw, or heatwave cease to be a background condition and become a logistics crisis, a budget crisis, a migration crisis, or a legitimacy crisis? Precision matters here. Deterrence offers little purchase against a feedback loop. Retaliation does not harden a runway. Military metaphors can lend drama while obscuring the civil, fiscal, and administrative work that actually determines whether societies cope or crack.

The harder and more sober view is that capability under climatic stress is composed of humble things. It is measured in crews available when roads close, in warehouses stocked before storms, in contracts that can surge without scandal, in standards revised before the next failure rather than after it, in backup communications that function when primary systems drop, in evacuation plans that account for care homes and dialysis patients rather than only healthy motorists. Trust, in this setting, is not an attitude survey. It is visible behaviour: whether people heed warnings, whether agencies share data without delay, whether mayors believe national ministries will answer the phone, whether after-action reports are candid because candour is rewarded instead of punished. Legitimacy has a ledger. It also has a tempo.

Migration adds another layer of pressure. Human mobility is among the oldest adaptive responses to changing conditions, and it should not be reduced to a security slogan. Yet receiving systems can be strained by sudden inflows even when movement reduces harm for households. Housing, school places, clinics, documentation, transport, sanitation, flood exposure on the urban edge: these are the points where abstract discussions about displacement become questions of administrative bandwidth. To describe that plainly is not to frame migrants as threats. It is to recognise that state capacity is tested at the level of procedures and services long before it appears in strategic doctrine.

The same is true of budgets. Political systems are skilled at funding visible response and chronically poor at rewarding quiet prevention. Temporary repairs are easier to defend than expensive redesigns whose success will be measured by the absence of catastrophe. Discounting can make long-horizon risks look manageable on paper even while short planning windows are closing in practice. A damaged culvert can be patched this year and next; a relocation or redesign invites headlines, scrutiny, and accusations of waste. Public narratives matter because fiscal legitimacy matters. When governments cannot explain why adaptation spending is prudent before the disaster, they are often forced to explain why the disaster was not foreseen after it.

This is why the central question is neither purely environmental nor narrowly military. It is diagnostic. Which assets are assumed to be reliable but no longer are? Which institutions carry the load when climate stress compounds? Where do decision rights sit when monitoring detects a problem but no one is clearly authorised to trigger action? Which standards still encode a climate that has already receded? Which services express sovereignty in practice, especially at the edge of territory? And where is the hairline crack already visible—in an insurance premium, a maintenance backlog, a delayed fuel barge, an evacuation route that functions only if three other systems remain intact?

No serious account of methane-related feedback risk can pretend to dissolve uncertainty. That would be false confidence dressed as clarity. The more honest task is to narrow the space in which uncertainty becomes an alibi for drift. There are futures that cannot be forecast with precision and choices that can still be judged in the present. Institutions can be designed to learn or designed to deny. Budgets can protect optionality or consume it. Infrastructure can be treated as an intergenerational promise or as a temporary convenience. Scientific cooperation can be preserved as a safety function even under geopolitical rivalry, or it can be allowed to erode until misperception becomes one more hazard.

The stakes are therefore not only climatic. They are constitutional in the small-c sense: they concern what citizens come to expect from public authority, and what public authority can still credibly promise when the physical substrate of planning is shifting. Strategic competition in the Arctic, the politics of standards and codes, the fragility of coastal and offshore systems, the dependence of critical infrastructure on hidden logistics, the difficulty of speaking honestly about tail risks without provoking paralysis or contempt—these are not separate conversations. They are facets of a single governing problem. The atmosphere sets conditions; institutions decide consequences.

The decisive question is not whether uncertainty can be abolished. It cannot. The decisive question is what a state chooses to count, test, and protect when the old ledger no longer captures the full exposure. If methane is treated only as a technical emissions issue, governments may continue to misread failures of reliability as isolated surprises. If it is treated as a strategic variable, the picture changes. Budgets, depots, ports, shelters, runways, mutual aid agreements, insurance markets, data rights, local knowledge, and institutional trust begin to appear in the same frame. That frame does not promise comfort. It offers something better: a way to judge whether a society is preparing to absorb shocks, or merely waiting for the next hairline crack to widen.

Chapter 1

Methane Risk in Security Planning

What kind of threat grows more serious precisely when institutions ask for clearer proof before they move? Security planning has long preferred dangers that can be counted, mapped, and assigned to a timetable. Methane unsettles that habit. Its significance lies not only in its physical effect but also in the way it compresses time for decision-making while leaving much of the surrounding uncertainty stubbornly intact. A ministry can price fuel, inspect a bridge, or draft an emergency protocol. It cannot command the pace at which background conditions lose their old reliability.

That mismatch is where the chapter begins: not with spectacle, but with administration. A road built on assumptions about seasonal cold, a port financed on assumptions about operating conditions, a pipeline maintained on assumptions about ground stability, a housing stock insured on assumptions about recurring risk - each looks manageable when treated as a discrete asset. The trouble begins when those assumptions start to move together. Then the ordinary machinery of government starts to strain. Procurement slows where conditions speed up. Maintenance cycles lengthen where windows for repair narrow. Budgets designed to smooth routine variation are forced to absorb changes that no longer behave like routine variation at all. The ledger remains on the desk. The ground beneath it shifts.

Security institutions do not fail only when they face overwhelming force. They also fail when their categories lag behind reality. Methane risk exposes that quieter danger. It resists the comforting fiction that climate harm can be filed neatly under environment while strategy concerns itself elsewhere. Once warming pressure begins to alter the dependability of land, water, transport, storage, and emergency access, the line between environmental and strategic issues ceases to be analytically useful. The question is no longer whether a climate process belongs inside security planning. The question is whether security planning can remain credible when it refuses to reckon with processes that erode the reliability of the systems it is meant to protect.

That makes timing the central tension. Long-run climate goals remain essential, yet governments do not govern only in the long run. They authorise projects, sign contracts, issue permits, set standards, and defend budgets under tighter deadlines. Methane sharpens pressure on those nearer horizons because it matters to the pace at which conditions can become less forgiving. This is not a moral ranking of gases, nor an invitation to treat one problem as a substitute for another. It is a matter of planning exposure. States inherit roads, depots, settlements, and supply systems that were not built for abstract futures but for particular expectations about stability. When those expectations weaken faster than institutions revise their decisions, strategy becomes reactive without admitting it.

The stakes are therefore larger than technical risk management. A state earns confidence through repeated demonstrations of competence in mundane tasks: keeping routes open, moving fuel, maintaining communications, reaching remote communities, repairing damage before it leads to abandonment. People do not experience state capacity as a theory. They experience it as whether the runway holds, whether the shelves are stocked, whether the warning arrives in time, whether the repair crew comes back before the season closes. When methane-related feedback risk is treated as distant or speculative, the practical consequence is often a dangerous delay in recognising how quickly small failures can accumulate into fiscal strain, political grievance, and loss of institutional trust.

And yet the subject punishes simplification. Feedback is not a licence for melodrama. Non-linearity does not mean every trend hides a sudden collapse, and uncertainty does not excuse institutional passivity. The discipline required here is harder than an alarm or dismissal. It asks decision-makers to hold two truths at once: some outcomes cannot be assigned tidy probabilities, and some preparations can still be judged prudent without that false precision. This is where the chapter places its weight. Not on theatrical claims about imminent breakdown, but on the quieter and more consequential problem of how public institutions behave when they know enough to worry, not enough to predict, and are still expected to answer for what they did or failed to do.

The path ahead follows that tension as it tightens. The opening movement turns to the issue of time, where near-term warming pressure collides with the long lives of infrastructure and capital commitments. From there the argument moves into feedbacks and forms of disruption that do not respect linear planning habits, before entering the harder terrain of deep uncertainty, where responsibility survives even when confidence intervals do not settle the matter. It then arrives at the strategic frame that gives the rest of the book its grip: exposure, sensitivity, and response capacity as the terms through which physical change becomes a question of state performance.

This framing matters because hazards do not govern themselves. Two jurisdictions may face similar physical pressures and produce very different outcomes because one has slack in logistics, room in its budget, clear lines of authority, and habits of institutional learning, while the other runs close to the margin even in calm conditions. The same thaw, the same fire weather, the same transport disruption can therefore register as an inconvenience in one place and a political crisis in another. Security analysis earns its keep when it explains that difference without mystification. Methane enters the argument here not as a symbol, but as a strategic variable that changes how quickly margins can disappear and how costly it becomes to pretend they have not.

A useful test sits quietly beneath all of this. When officials say they are prepared, what exactly are they claiming? Prepared for a known hazard on a familiar timetable, or for a shifting background in which assets age differently, seasons behave differently, and multiple systems come under pressure at once? A paper plan may satisfy administrative ritual. A contingency fund may reassure for a while. Neither answers the harder question of whether institutions built to manage forecastable strain can absorb disruption that changes the conditions of planning itself.

That is the friction that should remain in view before any narrower discussion begins. If methane alters not only climate outcomes but the speed and uncertainty with which they press upon infrastructure, logistics, and public authority, then the real issue is not whether governments can predict every turn. They cannot. The issue is what kind of state is revealed when the old assumptions fail: one that recognises the shortening window and adjusts before the crack spreads, or one that keeps reading from yesterday's map while the road has already started to give.

Why methane changes the timeline

In May 2020, a diesel storage tank at a power plant near Norilsk, in the Russian Arctic, collapsed and released fuel into rivers and tundra. The immediate failure was industrial, local, and brutally concrete: twisted metal, dark water, a spreading slick. Yet the event travelled quickly into the language of national vulnerability because the suspected cause pointed beneath the structure itself, to thawing ground in a region where frozen ground had long been treated as a dependable foundation. That is where the argument about methane begins. Not with a molecule in the abstract, but with a calendar that no longer holds still.

The point is easy to miss if climate change is imagined only as a slow rise in averages. States do not govern averages. They govern timetables. They approve housing projects meant to last decades, pipe fuel across difficult terrain, finance ports on the assumption of workable access, and stock remote communities according to seasonal rhythms that once felt dependable enough to budget. Methane matters here because its warming influence is concentrated in the near term compared with carbon dioxide, even though carbon dioxide remains decisive for the long-run climate outcome. That does not make methane morally worse, or politically more convenient, or scientifically more interesting than every other part of the problem. It makes it temporarily sharper. The issue is pace.

That sharper pace creates planning pressure because public systems are built on mismatched clocks. Atmospheric change can accelerate over years or a few decades in ways that matter to infrastructure managers, insurers, transport planners, and emergency officials. Ports, pipelines, roads, housing estates, substations, and water systems are financed, designed, and regulated over much longer horizons. A ministry can revise a briefing note in a week. It cannot rebuild a coastal depot or move a settlement at that speed. Once the nearer-term warming effect of methane is taken seriously, the governing question shifts. The question is no longer only how much change a system might face by the end of the century. It is how much confidence can still be placed in assumptions that support decisions being made now.

This is why timing belongs at the centre of security planning rather than in a technical annex. Security institutions often prefer threats that declare themselves in measurable inventories: troop numbers, fuel reserves, flood walls, response times, days of food, hours of backup power. Methane unsettles that habit because it changes the conditions under which those inventories retain meaning. A runway may still exist on paper, a winter road may still appear in the annual plan, a fuel delivery contract may still be signed, and a housing stock may still be booked as an asset. But if the season that makes these systems workable narrows or grows erratic, the formal existence of the asset tells planners less than it once did. The calendar starts to fray.

The practical effect appears first in ordinary decisions. A procurement officer wonders whether to order materials for a repair that can only be completed during a reliable cold period. A regional government delays capital work because ground conditions are uncertain, then pays more later when the problem worsens. An insurer raises premiums or narrows coverage, not because a grand theory has changed, but because repeated damage has altered the working judgement of risk. A municipal engineer stops asking whether an asset can be built to code and starts asking whether the code still describes the climate for which the asset is being built. None of these actions makes headlines on its own. Together, they change the tempo of government.

Methane's importance in this setting follows from a simple distinction that is often blurred in public debate. Long-lived greenhouse gases determine much of the long-run destination. Shorter-lived gases can exert disproportionate influence on the route and the speed at which dangerous conditions accumulate. For strategic planning, route and speed are not minor details. They shape when maintenance backlogs become intolerable, when emergency plans move from hypothetical to routine use, when housing becomes uninsurable, when remote access shifts from seasonal inconvenience to chronic uncertainty, and when treasury departments are forced to recognise that repeated patching is no longer cheaper than redesign. Stability is not only about end states. It is about the rate at which assumptions lose value before institutions have replaced them.

A useful comparison sits in Arctic and sub-Arctic logistics. Winter roads are not romantic relics. They are fiscal devices, supply corridors, and governance instruments. In parts of Canada, including the Northwest Territories, seasonal ice roads have long served mines, communities, and bulk freight movements that would be vastly more expensive by air. When such roads open later, close earlier, or operate under tighter load restrictions because conditions are poor, the effect is not confined to transport. Capital planning changes. Construction schedules slip. Fuel deliveries must be rearranged. Food costs can rise. A project that looked economical under one seasonal pattern looks fragile under another. The road itself becomes a question mark, and with it the decisions stacked behind it.

The Tibbitt to Contwoyto Winter Road, which serves diamond mines in the Northwest Territories, offers a clear illustration of how narrow margins matter. Its operational value depends on cold that is sufficient, sustained, and predictable enough to support heavy loads across frozen lakes and land sections. When a season is shortened or conditions are more variable, operators and customers do not merely tolerate inconvenience. They must recalculate stockpiles, freight priorities, maintenance windows, and capital expenditures. Here, the concept to name is the decision window. The phrase sounds technical, but the underlying reality is visible: fewer workable days for the same logistical task. That compression can force choices months earlier in the planning cycle and at a higher cost. A climate signal thus becomes an administrative deadline.

The same logic extends to assets that appear far more durable. Pipelines and storage facilities are commonly described as long-lived infrastructure, and rightly so. But long life is not the same as fixed reliability. The Norilsk spill in 2020 became emblematic not simply because a tank failed, but because it exposed a dangerous habit of thought: the belief that a frozen substrate could continue to function as an unspoken constant while warming progressed around it. Once that constant weakens, the timeline for inspection, reinforcement, redesign, and liability assessment shifts. The strategic implication is severe. Governments and firms may still be arguing about future scenarios while the maintenance calendar has already changed beneath them.

Housing reveals the same dynamic at a different social scale. Homes are political objects as much as physical structures. They anchor tax bases, local services, insurance markets, electoral promises, and family strategies. In regions affected by permafrost thaw, coastal erosion, repeated flooding, or unstable freeze-thaw cycles, housing risk does not wait politely for a national strategy document. It appears in cracked foundations, warped roads, burst pipes, repeated claims, and hard conversations about whether repair still counts as stewardship or has become denial. Once households, lenders, and local officials begin to act on different timelines, the state is pulled into a harder position. Either it revises its assumptions and bears visible costs sooner, or it delays and bears higher costs later under conditions of lower trust.

Trust matters here, but only if it is made concrete. In public life, trust means people believe warnings when they are issued, contractors believe payments will arrive, local officials believe emergency declarations will unlock support, and insurers believe standards are enforced rather than merely announced. Methane's near-term influence enters this field by straining the reliability on which such behaviours rest. If a community cannot be confident that the road will carry fuel when planned, it stocks more in advance where possible, pays more for alternatives where necessary, and presses governments harder on guarantees. If officials repeatedly communicate confidence in seasonal access and are then proved wrong, credibility drains faster than any press office can restore it. The warming effect itself may be atmospheric. The loss of confidence is administrative.

And yet this argument has a real limitation, and it deserves to be faced plainly. Not every infrastructure failure under warming conditions should be attributed to methane, or even to climate change alone. Some roads fail because they were poorly built. Some pipelines deteriorate because maintenance was deferred. Some settlements were sited in exposed places for reasons that had little to do with present climatic shifts. Historical underinvestment, weak governance, corruption, and simple engineering error remain powerful causes of disruption. The discipline required is to avoid turning methane into a universal explanation. It is better understood as a timing force that interacts with pre-existing fragilities. Where governance is competent, standards are updated, and slack exists in logistics and finance, systems can absorb more stress. Where institutions already run close to failure, even modest acceleration in physical change can have outsized effects.

That qualification does not weaken the case. It sharpens it. Strategic analysis is most useful when it distinguishes between hazard and exposure, between physical change and institutional consequence. Methane does not determine political outcomes by itself. It changes the pace at which governments must recognise that older planning assumptions are expiring. This is why the usual language of gradual adaptation often misleads. Gradualism suggests a generous sequence: observe, study, debate, revise, fund, build. In practice, methane-related warming pressure can shorten the interval between observation and consequence. By the time a formal review is complete, the system under review may already be facing more demanding conditions than the review assumed. The calendar frays again.

A second named case helps show how wide this problem can spread. In Alaska, the Dalton Highway and associated North Slope infrastructure have long required engineers to reckon with permafrost, frost heave, drainage, and shifting ground conditions. The road exists because governments and firms accepted that Arctic logistics demanded continuous adaptation rather than one-off construction. That is the lesson worth keeping. Some institutions do adjust. They monitor, compare, reinforce, reroute, and accept higher maintenance burdens as the price of continued function. The existence of such adaptation does not cancel the strategic problem. It shows what the problem costs when taken seriously. It also reveals a harder truth: the places that adapt best are often those with unusual financial resources, concentrated strategic importance, or both. Many municipalities and regional systems do not enjoy those advantages.

The moral temptation in climate politics is to turn every discussion into a ranking of villains and virtues. That temptation is understandable and often politically useful. It is less helpful here. Security planning has to decide under constraint, and constraint is usually temporal before it is philosophical. Which assets require redesign sooner than expected? Which supply systems need greater slack because reliable access windows are narrowing? Which procurement rules slow action at the very moment shorter decision windows demand earlier commitment? Which housing strategies still assume continuity where continuity has become expensive to defend? These are not questions of atmospheric purity. They are questions of statecraft.

Seen in this light, methane changes the timeline because it narrows the distance between climate signal and institutional consequence. That narrowing is what makes it strategically important. It does not replace the long-run centrality of carbon dioxide. It does not permit sensational claims about imminent collapse every time a season turns strange. It does demand a more exact form of seriousness: one that recognises how quickly a physical process can become a budgeting problem, a logistics problem, an insurance problem, a service-delivery problem, and then a political problem. The old calendar assumed more time between warning and redesign. That assumption is losing value.

The hardest question follows from that loss. When officials say they are planning for stability, which clock are they using: the lifespan of the asset, the cadence of the budget, the rhythm of the election, or the pace at which the conditions beneath all three are shifting? Until those clocks are brought into the same frame, the danger will not simply be warmer air. It will be a state making long-lived promises on a shortening timeline.

Feedbacks and non-linear disruption

In May 2020, a fuel tank at a power plant near Norilsk collapsed and released diesel into Arctic waterways. The operator was NTEK. The place was northern Siberia. The date is easy to check. What made the event travel so widely through policy circles was not only the scale of the spill, but the suspicion that the ground beneath the structure had become less stable as permafrost thawed. A piece of industrial equipment failed, yet the more unsettling lesson sat one layer lower: a foundation once treated as background had become active. The hinge had moved.

That is the essential character of feedback risk. It does not merely add more of the same pressure. It changes the relationship between pressures, surfaces, institutions, and time. Linear expectation assumes that if conditions worsen by a given amount, damage rises in roughly proportionate fashion. Feedbacks disturb that comforting geometry. They can accelerate change by reinforcing the very processes that produced the change in the first place, or by shifting a system across a threshold after which familiar rules no longer describe what happens next. Small increments can matter for years. Then the pattern breaks.

For security planning, that distinction is decisive. A gradual trend invites conventional management. Budgets can be stretched. Maintenance can be phased. Procurement can follow ordinary cycles. Threshold-like behaviour is different. It creates moments when yesterday's rate of change is a poor guide to tomorrow's operational demand. A road remains usable until it is not. A fire season looks difficult until evacuation routes fail. A supply chain runs lean until one missing segment forces the whole system into delay. Non-linearity, in this sense, is not a mystical property of nature. It is a practical warning that some systems absorb pressure smoothly while others absorb it badly and then suddenly.

Climate feedbacks make this harder because they combine physical acceleration with imperfect attribution. Warmer conditions can increase fire risk, alter soil moisture, weaken ice, or affect permafrost stability. In some cases, warming can, in turn, amplify methane release from wetlands or thawing ground, further intensifying near-term warming. The science here requires care. Feedback language can describe mechanisms supported by evidence without granting a licence to treat every extreme event as proof of runaway change. Mechanism matters. So does restraint. The serious claim is not that all climate systems are poised on a knife-edge or that any unusual season reveals a hidden tipping point. The serious claim is narrower and more demanding: institutions built around linear assumptions can be surprised by systems that do not degrade in neat proportions.

The difference becomes visible when one compares a trend with a break. Sea level rise, for all its complexity, often enters planning as a long, accumulating pressure. A port authority can model elevations, revise design assumptions, and stage investment over time, even if those choices remain politically difficult. By contrast, the loss of road access during a season when planners expected one more year of tolerable conditions can trigger a cluster of immediate decisions: emergency procurement, alternate freight, deferred construction, revised insurance, public messaging, and sometimes political blame. The physical change may not have been larger in absolute terms than a previous year's shift. What changed was the operational state of the system. The hinge moved again.

The language of thresholds is useful here, provided it is not abused. A threshold does not mean an apocalyptic cliff. More often it marks a point at which a system changes regime for planning purposes. Frozen ground that once provided predictable support becomes variable enough that inspection and repair schedules no longer hold. Heat that once strained emergency departments becomes intense enough to simultaneously alter labour routines, transport performance, and electricity demand. A communication network that functioned with modest redundancy becomes brittle when several dependent services require it under concurrent stress. The threshold is administrative as much as physical. It is the point at which ordinary procedures stop matching reality.

British Columbia offered a stark reminder in 2021. The June heat dome brought extraordinary temperatures, including records in places such as Lytton, which was later devastated by fire. Months later, in November 2021, severe flooding and landslides cut major highways and disrupted rail links. These were not methane events in any simple sense, and serious analysis should resist that shortcut. They are relevant because they showed how systems planned through separate hazard files can become non-linear in combination. Heat stress, fire damage, transport interruption, supply delays, emergency shelter, agricultural losses, and public communication did not line up politely. They overlapped. Procurement and emergency systems were forced to cope not with a single event but with altered operating conditions across several fronts. The result looked sudden. Much of it was structurally primed.

That phrase matters: structurally primed. Non-linearity does not mean that surprises come from nowhere. Usually they emerge from a system already thinned by efficiency pressures, fragmented mandates, ageing assets, or weak buffers. This is where governance enters the argument with full force. The natural sciences can identify mechanisms, estimate ranges, compare scenarios, and warn against simplistic inference. They cannot decide how much spare capacity a fuel network should retain, how fast a building code should be revised, when a mutual aid agreement ceases to be credible, or whether a ministry is authorised to act on uncertain evidence. Those are governance questions. They determine whether non-linear physical change becomes a manageable strain or a political crisis.

The problem is easy to misstate. Some policy rhetoric treats non-linearity as if it automatically justifies maximum alarm and unlimited spending. That is not serious. Public institutions must choose under constraint, and much of what they face remains uncertain in timing, scale, and interaction. Yet the opposite error is more common in administrative practice: demands for a level of predictive precision that no complex system can honestly supply. A procurement rule asks for firm thresholds before funds can be released. A treasury asks for quantified probabilities before a contingency stock can be expanded. A local authority waits for a formal designation before changing maintenance practice. Each demand sounds prudent in isolation. Together, they can create delay precisely where uncertainty argues for earlier, more flexible preparation.

A second named example shows the pattern from another angle. In August 2023, authorities in the Northwest Territories ordered the evacuation of Yellowknife as a wildfire threatened the city. The event was notable not only because of the fire behaviour. It revealed how quickly several planning assumptions can narrow at once in a remote jurisdiction: road capacity, airlift availability, accommodation, continuity of care, fuel supply, public communication, and the management of fear. Emergency systems rarely fail because one official lacks a map. They fail when dependencies tighten faster than decisions do. What looked like a fire problem soon became a logistics problem, a health services problem, and a credibility problem. Non-linearity here did not mean ignorance of risk. It meant that the coupling between systems produced consequences out of proportion to any single variable on its own.

Trust, often invoked too casually, becomes concrete in such moments. It is visible when residents leave early enough for evacuation routes to function, when hospitals transfer patients before transport bottlenecks harden, when contractors accept emergency work because they expect payment and clear command, when neighbouring jurisdictions believe shared plans will hold under load. These are not sentiments. They are behaviours. Feedback-driven disruption can erode them because repeated surprises teach institutions and publics a costly lesson: that official timelines no longer match lived reality. Once that gap widens, every later warning becomes harder to deliver and every later reassurance harder to believe.

Still, a caution is necessary. Not every abrupt disruption is evidence of climatic non-linearity, and not every feedback process produces threshold-like governance effects. Some systems remain broadly linear over the timescales that matter for particular decisions. Some infrastructure performs well precisely because engineers planned for uncertainty, monitored intensively, and accepted high maintenance costs. Parts of Arctic energy infrastructure, for example, have long been designed around difficult ground conditions rather than naive assumptions of permanence. That matters because it prevents the analysis from becoming theatrical. Non-linearity is not a synonym for inevitable collapse. It is a warning that rates of change, system interactions, and hidden dependencies matter more than linear budgeting habits tend to admit.

The task, then, is not to turn every risk register into a catalogue of worst cases. It is to distinguish where gradualism remains a workable planning model from where it becomes an excuse for institutional lag. A seawall project may still lend itself to staged adjustment. A winter road network dependent on narrow cold windows may not. A standard procurement cycle may suit routine equipment replacement. It may fail badly when repair seasons shrink or concurrent emergencies exhaust suppliers across a region. The analytical move is simple, though its implications are not: ask which systems degrade smoothly, which can flip into a different operating state, and which public rules assume the former when they are plainly governing the latter.

Methane sharpens this issue because it belongs to the part of the climate problem where timing and acceleration carry unusual weight. Its importance in this discussion is not symbolic. It lies in how feedback-related warming pressure can help push certain systems from trend into disruption before long-lived institutions have adjusted their habits. When that happens, the most dangerous error is often conceptual before it is technical. Officials continue to treat a changed operating environment as a string of exceptions. Budgets classify repeated emergency spending as temporary. Maintenance teams describe recurring deformation as unusual. Communications staff call compounding strain unprecedented one year after calling something else unprecedented the year before. The hinge moves, but the paperwork does not.

What feedback language can properly claim is therefore both less and more than public argument usually allows. It can be claimed that some processes reinforce warming and can accelerate underlying pressures. It can be claimed that thresholds exist in physical and social systems, though their exact location may be uncertain. It can be claimed that linear extrapolation often misleads when systems are tightly coupled. It cannot honestly claim that every unusual event proves a runaway dynamic, that all tipping elements are imminent, or that uncertainty disappears once the word feedback has been invoked. Mature analysis keeps mechanism and humility together.

That combination is harder than it sounds because political institutions are often organised to prefer either confidence or silence. Confidence reassures until it fails. Silence protects reputations until events overtake it. Feedback-driven disruption punishes both habits. It requires officials to communicate that some hazards may intensify in uneven ways, that some thresholds are known only after they have been crossed, and that prudent action may rest on vulnerability rather than forecast certainty. This is why non-linearity belongs in governance analysis, not only in scientific briefings. It changes what counts as adequate preparation, what counts as evidence, and what counts as delay.

The deeper question is uncomfortable because it reaches beyond climate modelling into the design of public authority itself. When a system may not change in neat increments, how much flexibility should institutions preserve in budgets, contracts, staffing, and decision rights before that flexibility looks inefficient? Keep too little, and the next threshold becomes a crisis of capability. Keep too much, and administrations will be accused of paying for contingencies that never quite arrive on schedule. That is the real pressure point. The hinge is not only in the ice, the soil, or the atmosphere. It is in the state's willingness to govern for breaks in pattern before the break has been fully priced.

Deep uncertainty and state responsibility

In March 2011, officials at the Fukushima Daiichi nuclear power station were not confronting a shortage of data. They faced a flood of it: seismic readings, loss-of-power reports, failing cooling systems, fragmentary field observations, contradictory forecasts, and a public that needed answers before certainty was possible. The operator was Tokyo Electric Power Company. The place was Fukushima Prefecture. The date is fixed. What made the disaster so instructive for any discussion of deep uncertainty was not that experts knew nothing. It was that they knew enough to act, too little to predict with confidence, and still had to make decisions whose consequences would be judged as though a single clear picture had existed from the start.

That tension is closer to climate security planning than many administrations like to admit. Public institutions are often organised around a quiet promise: if risk can be measured well enough, the state can provide the right level of protection. Budgets, codes, emergency plans, and insurance arrangements are all shaped by that promise. It is a reasonable habit when probability distributions are stable, historical baselines remain useful, and the next disturbance resembles the last. Deep uncertainty breaks that contract. The map blurs.

Deep uncertainty does not mean total ignorance. It means decision-makers lack one trusted probability distribution around which to organise action. They may know the mechanisms, the direction of travel, the range of plausible harms, and the fragilities already visible in assets and institutions. What they do not possess is the kind of settled forecast that allows a treasury, a ministry, or a regulator to say with confidence: this is the level of risk, this is the correct design threshold, this is the right amount of spare capacity, this is the exact moment at which the old standard expires. For methane-related climate feedback risk, that gap matters because timing, interaction, and threshold effects resist the tidy arithmetic on which much administrative accountability depends. Not enough certainty exists for comfort. Enough exists for responsibility.

The usual administrative response is to reach harder for precision. Committees ask for narrower numbers, sharper thresholds, cleaner attribution, stronger confidence statements. A funding request is returned because the scenario range is too wide. A maintenance upgrade is delayed because the trigger point has not been formally crossed. An emergency stockpile remains unchanged because no single model commands sufficient authority. Each decision can be defended in isolation. Together they can amount to a style of failure. False precision does not simply misdescribe reality; it trains institutions to act too late. The file stays open.

This is the core dilemma. A predict-and-provide model assumes that once a hazard is estimated, the state can build the corresponding protection and then manage exceptions. That model has deep roots in modern governance. Flood defences are set to the expected return periods. Roads are engineered to design conditions. Emergency services plan against defined scenarios. Welfare systems estimate demand categories. The virtues are obvious: comparability, procedural clarity, budget discipline, and visible lines of accountability. Yet the model performs badly when the hazard itself is shifting, when compounding pressures matter more than isolated events, and when the cost of under-preparation is not merely material damage but loss of institutional credibility.

A different approach begins from a less flattering but more realistic premise: some decisions must work tolerably well across a range of futures that cannot be ranked with confidence. That is the logic of robustness. It does not ask what the perfect plan is for the most likely future. It asks what kinds of plans fail least badly across several plausible futures, including ones that arrive earlier, combine more hazards, or expose dependencies that routine planning ignored. In climate governance, this often takes the form of adaptive planning, staged commitments, trigger points, modular investment, and deliberate preservation of slack. The point is not elegance. It is survivability.

The distinction becomes easier to see in infrastructure decisions. Imagine a northern jurisdiction deciding whether to reinforce a fuel depot, redesign a drainage system, or relocate housing from unstable ground. A predict-and-provide mindset asks for the best estimate of future conditions and then tries to optimise against that estimate. A robustness mindset asks a harder question: if future conditions diverge from the estimate in several plausible ways, which option preserves the most room to adjust without catastrophic regret? The first approach flatters administrative habits because it yields a single number, a preferred option, a clean business case. The second often looks untidy. It may recommend spending more now to keep choices open later, or building in stages, or accepting lower efficiency in ordinary years in exchange for less brittle performance under stress. That can sound indecisive. Often it is the opposite.

The problem is not confined to capital projects. It reaches into procurement, staffing, logistics, and emergency authority. A health service that runs at near-full occupancy may look efficient in annual reporting and still fail under compound disruption because it lacks surge room. A supply network that cuts warehousing to save money may work until a road, port, or rail link fails during a simultaneous regional emergency. A mutual aid arrangement may appear prudent until every neighbouring jurisdiction needs the same assistance at once. In each case the institution did not fail because no one knew hazards existed. It failed because planning rewarded a narrow definition of efficiency and punished capacity that could not justify itself through routine utilisation. Unused slack is always vulnerable in a spreadsheet. It becomes priceless in a crisis.

One reason deep uncertainty is so politically awkward is that democratic systems demand reasons that sound exact. Ministers must defend expenditure before legislatures. Local authorities must justify code changes to residents and developers. Treasury officials must compare one risk claim against another. Insurers must defend pricing decisions. Courts may ask whether a decision was proportionate to the evidence available at the time. These are not trivial demands. They protect public money and restrain arbitrary power. But they also create a bias towards numbers that appear firmer than the underlying reality permits. The result is a peculiar ritual of certainty: models presented with more confidence than modellers feel, thresholds treated as objective when they are partly conventional, adaptation promises made on the assumption that future revisions can always be managed later. The language sounds disciplined. The practice can be evasive.

A named case from the United Kingdom helps expose the pattern. During the COVID-19 pandemic, public attention often settled on the phrase "following the science", as though science could supply a single deterministic script for policy under fast-moving uncertainty. It could not. Scientific advice clarified mechanisms, ranges, and trade-offs; political institutions still had to decide under incomplete information about timing, compliance, capacity, and unintended effects. The relevance here is not that a pandemic and methane feedback risk are the same. They are not. The relevance lies in the administrative temptation. Under public pressure, institutions often translate uncertain evidence into overconfident rhetoric because they believe ambiguity will look like weakness. That move may buy temporary authority. It can also deepen mistrust when revision becomes unavoidable.

A second example comes from the Netherlands, a country rightly admired for flood management. Dutch water governance is often praised because it does not rely solely on one-off prediction. It combines engineering, long-term planning, periodic reassessment, and institutional arrangements designed to accommodate changing conditions. The details differ from climate feedback planning in important ways, and flood risk can often be modelled more specifically than methane-related cascading risk. Still, the example matters as a counterpoint. It shows that high-capacity states can design institutions that treat uncertainty not as a reason for paralysis but as a condition to be managed through iteration, redundancy, and visible commitment. The lesson is not that all uncertainty can be tamed by better engineering. It is that responsibility need not wait for perfect foresight.

That counter-example also marks the limitation of the argument. Robustness is not free. Adaptive planning can become an alibi for delay if governments endlessly defer hard choices under the banner of flexibility. Excess redundancy can waste scarce resources. Trigger points can be gamed or ignored. Modular design can produce mediocrity if no stage is ever completed properly. Not every institution needs to be built around worst-case logic, and not every uncertainty warrants expensive precaution. These are serious objections. They deserve more than a gesture. The answer lies in distinguishing between domains where reversibility is high and domains where it is not. A procurement contract for replaceable equipment can tolerate more error than a settlement pattern, a major pipeline corridor, or a coastal asset designed to last half a century. Where decisions are sticky, consequences cascade, and correction arrives slowly, the burden of proof should shift. The file should not remain open merely because the forecast is imperfect.

State responsibility under deep uncertainty therefore depends less on omniscience than on design. A capable institution makes its uncertainty legible. It states what is known, what is contested, what assumptions are carrying the plan, and what signal would trigger revision. It defines authority before the emergency rather than during it. It preserves enough slack in finance, logistics, and staffing to absorb being wrong in one direction without collapsing in the other. It builds reporting systems that reward early warning rather than punishing unwelcome evidence. It does not confuse decisiveness with inflexibility. These behaviours are visible. One can inspect whether budgets include contingency capacity, whether standards are revised on schedule, whether drills test simultaneous failure, whether after-action reports alter procurement rules, whether local knowledge is permitted to modify central plans. Responsibility here is not a mood. It is an institutional practice.

This matters especially for methane-related feedback risk because the pressure does not present itself as a single, schedulable event. It acts through timing, interaction, and background instability. One year may produce unusual thaw, another transport disruption, another a sequence of fire and smoke events that strains health systems and evacuation capacity. Waiting for one agreed distribution of risk before altering capital planning or service design is therefore a recipe for perpetual lateness. By the time the distribution looks settled enough to satisfy every procedural demand, the conditions that matter on the ground may already have moved. Deep uncertainty becomes a bureaucratic loop: more evidence requested, more time consumed, fewer options left.

The file stays open because institutional culture often mistakes revision for failure. Plans are written as if authority depends on certainty; when conditions change, officials fear that acknowledging change will damage credibility. Yet credibility often depends on the opposite. It depends on whether institutions announce in advance that uncertainty exists, identify what they are watching, and revise openly when triggers are met. A public that sees this pattern repeatedly may not enjoy the uncertainty, but it learns that adjustment is part of competence rather than proof of confusion. Trust then rests on visible habits: warnings updated without euphemism, spending justified through scenarios rather than false confidence, contingency stocks maintained even when unused, local disruptions logged and compared rather than dismissed as anomalies.

The deeper shift is philosophical but practical. Under deep uncertainty, the state cannot promise to predict correctly in every case. It can promise to remain governable when predictions fail. That is a lower claim and a more serious one. It reframes preparedness away from the fantasy of exact foresight and towards the architecture of correction. Can the institution absorb surprise without administrative panic? Can it change course without legal paralysis, fiscal collapse, or public disbelief? Can it protect essential functions while information remains incomplete? Those are strategic questions, not technical footnotes.

Methane feedback risk sharpens the stakes because it compresses time for institutions already tempted to overvalue certainty and undervalue optionality. If a government continues to demand a single decisive forecast before changing standards, budgets, or asset strategies, it may preserve procedural neatness while losing practical control. If it moves too loosely, it risks waste, confusion, and accusations of acting on speculation. That trade-off will not disappear. The useful question is not how to banish uncertainty from public decisions. It is how much imprecision a state can admit honestly while still committing early enough to prevent the file from becoming an archive of missed chances.

Methane as a strategic variable

In August 2023, authorities in the Northwest Territories ordered the evacuation of Yellowknife as wildfire threatened the city. The immediate hazard was fire. The relevant place was northern Canada. The date is easy to verify. Yet what made the episode strategically revealing was not only flame or smoke. It was the way a physical threat exposed the state through its dependencies: road access, airlift capacity, fuel supply, continuity of care, shelter arrangements, public communication, and the pace at which decisions could be coordinated across agencies. One hazard, many points of strain. The chain matters.

That chain is the right place to situate methane. Methane-related feedback risk should not be treated as a symbolic add-on to climate discussion, nor as a detached atmospheric curiosity. It becomes strategically meaningful when it alters the conditions under which public authority must keep systems functioning. A security analysis that begins and ends with emissions misses the state entirely. A security analysis that begins with state performance but ignores physical drivers misreads the timetable. The useful move is to place methane in the middle, as a strategic variable that helps explain why certain jurisdictions face sharper exposure, greater sensitivity, and weaker response under climatic stress than others facing broadly similar physical pressures.

This shift in framing matters because physical hazard alone does not determine political outcome. Two regions can face comparable warming pressures, comparable thaw, comparable fire weather, or comparable disruption to seasonal access and still produce very different results. One preserves continuity through redundancy, disciplined maintenance, clear decision rights, and fiscal room to absorb shocks. The other accumulates repair backlogs, relies on optimistic seasonal assumptions, narrows spare capacity in the name of efficiency, and discovers too late that multiple systems were running closer to failure than official reporting suggested. The atmosphere does not choose between these outcomes. Institutions do.

Exposure is the first term in the method, and it should be understood plainly. Exposure asks what lies in the path of hazard: which assets, which communities, which routes, which services, which budgets, which commitments. A thawing substrate means little strategically unless something valuable depends upon that substrate remaining stable. A shift in winter road reliability matters because food deliveries, mining schedules, fuel stocks, construction plans, and household costs may all depend upon a narrow cold season. Exposure is visible. One can map roads, pipelines, depots, settlements, airstrips, substations, housing, and communications lines. One can inspect whether a hospital depends upon a single road corridor, whether a remote community relies on a short shipping window, whether storage tanks stand on ground now showing signs of instability. Exposure is not an attitude. It is a placement.

The 2020 diesel spill near Norilsk remains a powerful example because it exposed more than industrial negligence or local fragility. A tank failed at a site whose basic security depended on the reliability of frozen ground in the Russian Arctic. The hazard was not merely that temperatures were changing; it was that critical infrastructure had been placed within a system whose old background assumptions were no longer dependable enough to remain unexamined. Exposure, in this sense, is not just proximity to nature. It is the degree to which a state has tied essential functions to environmental conditions that are becoming harder to trust. The strategic implication follows quickly: the more concentrated the dependency, the more severe the exposure when conditions move.

Yet exposure alone explains too little. Some systems are highly exposed and still cope tolerably well. Others are only moderately exposed and fail badly. That difference brings in the second term: sensitivity. Sensitivity asks how strongly a system reacts once hazard reaches it. A community with multiple access routes, reinforced storage, flexible procurement, and ample local authority may absorb a poor season with difficulty but without breakdown. A community relying on one winter road, thin inventories, fragile power infrastructure, and slow administrative escalation may experience the same physical disruption as a wider crisis. Sensitivity is the measure of how much damage a given level of stress produces. It is not reducible to geography. It is shaped by design, maintenance, dependency, and institutional habit.

Take seasonal access in northern Canada. Winter roads have long served communities and industrial operations by enabling bulk movement over frozen terrain during limited periods. When those periods shorten or become more variable, the exposed asset is obvious: the road itself and the supply chain attached to it. But sensitivity differs. A mining operation with capital reserves, alternative freight options, and contractual flexibility may suffer higher costs yet continue operating. A small settlement dependent on bulk fuel deliveries, affordable food transport, and public service routines built around that same access window faces something sharper. The physical hazard may be comparable. The institutional effect is not. Sensitivity turns climatic pressure into unequal political reality.

Sensitivity also reveals why abstract talk of resilience often conceals more than it clarifies. A system is not resilient because it has issued a strategy or run an exercise. It is resilient when strain does not produce cascading loss of function. That can be measured in ordinary ways: how quickly a damaged road is repaired, whether backup fuel is actually stored and rotated, whether a power interruption knocks out communications and water pumps, whether evacuation support includes medically vulnerable residents rather than assuming generic mobility, whether alternative suppliers can be activated without weeks of contractual delay. Once tied to visible behaviours and decisions, sensitivity becomes assessable rather than rhetorical. It moves from promise to performance.

The third term is response capacity, and here the analysis turns unmistakably strategic. Response capacity asks what a jurisdiction can actually do once stress arrives. It includes emergency services, transport, warehousing, finance, technical staff, legal authority, data systems, and political legitimacy. Most importantly, it includes how these elements fit together under pressure. A government may own aircraft and still lack evacuation capacity because crews are stretched, weather windows are poor, and receiving facilities are full. A ministry may hold monitoring data and still fail because no one has authority to trigger action before formal thresholds are crossed. Capacity is relational. It depends on coordination as much as equipment.

Yellowknife in 2023 illustrated this point with brutal clarity. The hazard was obvious enough. The strategic question was whether roads, flights, communications, care systems, and public messaging could be aligned fast enough to preserve order and safety under a narrowing window. Capacity did not reside in any single ministry or machine. It resided in the fit between them. When that fit holds, governments appear competent even amid danger. When it frays, citizens experience the state as delay, contradiction, or absence. Methane enters the frame because feedback-driven warming can increase the frequency or intensity of conditions that stress such systems, and because it can do so on timelines that expose every administrative weakness already built into them.

This three-part method matters because it disciplines argument. Climate debate often slips into vague claims that a given hazard will destabilise a region or threaten national security. Sometimes it will. Sometimes it will not, or not in the way predicted. Exposure, sensitivity, and response capacity force the analyst to ask harder questions. What exactly is exposed? How strongly does it react? Who can respond, with what resources, under which authority, and how quickly? Those questions reduce the temptation to confuse dramatic hazard with inevitable breakdown. They also reduce the temptation to treat stable conditions in one jurisdiction as proof that the hazard is overblown. Stability may reflect capacity, not safety.

A counter-example helps show the point. Parts of Arctic infrastructure in Alaska have long been engineered with a degree of seriousness that acknowledges unstable ground, difficult logistics, and costly maintenance as permanent features rather than unfortunate surprises. Elevated pipeline segments, intensive monitoring, and continual adjustment do not remove exposure. They change sensitivity and capacity. The system remains vulnerable to harsh conditions, but it does not respond as a neglected system would. This is a genuine limitation on any sweeping climate-security claim. It shows that exposure need not produce crisis if institutions accept high maintenance burdens, preserve technical competence, and build authority around adaptation rather than denial. The strategic lesson is not comforting, but it is clarifying: many failures attributed to nature are in part failures of preparation, funding, or governance design.

Still, that example should not be romanticised. High-capacity adaptation is expensive. It often depends on unusual strategic importance, concentrated revenue, or sustained political attention. Those conditions are not evenly distributed. A critical energy corridor, military asset, or nationally symbolic project may attract maintenance discipline that ordinary municipal infrastructure never receives. This is why fiscal resilience belongs inside response capacity rather than beside it. Money is not everything, but it changes the menu of possible responses. A government with reserves, borrowing room, and procurement flexibility can relocate assets, charter transport, stockpile essentials, or redesign infrastructure sooner. A government operating through austerity, fragmented funding streams, or repeated emergency appropriations tends to patch rather than rebuild. The same hazard then produces different institutional futures.

Administrative coherence matters just as much. Two agencies may both recognise rising risk and still fail collectively because mandates overlap, data cannot be shared quickly, or legal authority sits at the wrong level. Public institutions often report strength in vertical lines while the real weakness lies in the gaps between them. A transport ministry may know a road is degrading. A finance department may know the repair budget is unrealistic. A local authority may know community needs are changing. An emergency management office may know evacuation plans rely on assumptions already overtaken by recent events. If these insights do not converge into timely decision, methane-related feedback risk becomes another force exploiting organisational seams. Strategic weakness often hides in those seams.

Trust, so often invoked as a moral good, should be read here as an operating condition. It becomes visible when communities heed warnings, when officials share bad news early, when contractors mobilise under emergency terms without expecting bureaucratic chaos, when regional governments believe national funds will arrive, and when after-action reviews alter actual spending or standards. Methane as a strategic variable affects trust indirectly but powerfully. Repeated disruption teaches publics whether official assurances still map onto lived reality. If roads fail after being described as dependable, if fire seasons overwhelm systems that were publicly labelled prepared, if repeated repairs are announced as solutions and then repeated again, trust decays not because citizens have become irrational but because experience has revised their judgement.

This is why methane should be treated neither as a singular cause nor as a mere background factor. It is a strategic variable because it changes the tempo and severity of the pressures acting on exposed systems, which in turn reveals differences in sensitivity and response. That formulation is especially useful for case comparison. It allows an analyst to examine ports, pipelines, settlements, offshore platforms, health networks, or migration systems without assuming that each hazard operates identically. The categories remain stable while the particulars change. A coastal fuel depot and a remote housing settlement differ in material form, but both can be assessed through the same lens: what is exposed, how reactive the system is, and what capacity exists when conditions turn against it.

The method also guards against a common analytical error: treating security as a dramatic endpoint rather than a cumulative process. States rarely move from order to disorder in a single leap. More often they absorb one disruption, then another, then a third that lands on a system already weakened by the first two. Maintenance is deferred. Insurance tightens. Skilled staff leave. Communities lose confidence. Emergency spending becomes routine. Political arguments harden. By the time a crisis is named, much of the strategic change has already occurred through smaller administrative failures. Methane-related feedback risk belongs in this picture because it can intensify the background conditions under which such attrition accelerates. The state weakens by instalments.

Once viewed this way, the baseline method becomes less a formula than a discipline of attention. Exposure directs the eye to placement and dependency. Sensitivity asks how stress travels through design and interconnection. Response capacity examines the ability to act under time pressure, fiscal strain, and imperfect information. Together they link climate process to state performance without pretending that physics alone explains politics or that politics can bargain its way out of physical constraint. The method is spare by design. It resists grandiose claims. It also resists complacency.

Its usefulness lies in what it makes impossible to ignore. A government may reduce methane emissions and still remain strategically brittle if logistics are thin, fiscal slack is absent, and mandates are incoherent. Another may face rising climatic pressure yet preserve continuity for longer because it invests in maintenance, keeps redundancy where efficiency doctrine would prefer none, and allows monitoring to trigger action before crisis language becomes unavoidable. These are not abstract contrasts. They shape whether ports remain reliable, whether winter roads can still anchor supply planning, whether housing remains insurable, whether evacuation plans remain credible, and whether public authority appears present when conditions deteriorate.

The harder question follows immediately. If methane is treated as a strategic variable, not merely an environmental indicator, what should governments count as evidence that their institutions are ready: emissions inventories, hazard maps, reserve funds, spare lift capacity, revised standards, intact trust, or some less comfortable combination of all five? That question cannot be settled by atmospheric science alone. It has to be answered in budgets, depots, maintenance logs, emergency protocols, and the quality of the state's fit with the territory it claims to govern.

Chapter 2

Security Concepts for Climate Feedbacks

What makes a threat strategic when no enemy has chosen it, no border has been crossed, and no declaration has been issued? Security language was shaped by contests of will. It grew comfortable with adversaries, signals, deterrence, retaliation, and the visible grammar of intent. Climate feedbacks do not fit that grammar neatly. They can damage territory, budgets, supply systems, and public confidence without harbouring motive, and they can do so through chains of dependency that remain politically easy to misdescribe until failure has already begun. The familiar labels still sit on the folder. The contents do not.

This mismatch matters because language does more than describe danger; it arranges responsibility. Call a storm a hazard, and the term may suggest something regrettable but external, a blow from outside ordinary governance. Call the same event a security threat, and the language can sharpen attention but also invite the wrong analogies, as though coercion, deterrence, and retaliation were waiting just offstage. Between those two errors sits the harder work of concept. Precision is not an ornament here. It decides whether institutions look in the right place when systems begin to fail.

Imagine a coastal settlement after a severe storm. Roofs are torn, roads are blocked, and power lines are down. The damage is visible and immediate. Then another problem emerges: fuel cannot move. Emergency vehicles queue. Backup generators start to falter. Communications weaken. Shops ration what remains. The storm caused destruction, but the deeper strategic harm may come from the distribution failure that follows, from the way an interruption in one service reaches across many others. The queue becomes the motif of the problem. A queue for fuel, a queue for repairs, a queue for decisions. What appears natural at the front end becomes institutional at the back.

That shift from event to consequence marks the central tension. Security analysis needs enough breadth to take climate feedbacks seriously and enough discipline to avoid turning every disruption into a slogan. Loose language can mobilise attention for a season and still leave decision-makers with blurred categories, inflated claims, and muddled priorities. Tight language can feel slower, less dramatic, even cold. Yet it is the tighter vocabulary that makes comparison possible. Without it, hazard gets confused with threat, vulnerability with weakness, capability with budget size, and strategy with posture alone.

The stakes are practical long before they are theoretical. Governments allocate money, assign authority, revise standards, and justify emergency powers through concepts that appear dry until they fail. If those concepts are vague, then preparedness claims become vague as well. A ministry says the system is resilient, but means only that a plan exists. An agency says a risk has been mitigated, but means only that a warning has been issued. A government says a threat has multiplied, but cannot say through which institution, by which dependency, or at what point routine strain becomes strategic instability. Bad categories do not remain in speeches. They move into budgets, procurement rules, and after-action reports.

This also explains why the chapter must keep a careful distance from borrowed military habits of mind. Deterrence works, where it works at all, by shaping an opponent's calculation. Feedback-driven risks do not calculate. A thawing substrate is not dissuaded by resolve. A compounding emergency does not retreat because a state has signalled willingness to retaliate. Importing those logics into questions of preparedness can produce a dangerous theatricality: much talk of national will, too little scrutiny of warehouses, staffing, interoperability, mutual aid assumptions, and legal authority under stress. The queue lengthens while the rhetoric hardens.

Capability, then, needs stricter handling. It is not simply a synonym for more money, larger agencies, or more equipment on paper. Capability appears in whether logistics hold when several services need the same corridor, whether staff can make decisions without waiting for distant clearance, whether information moves between institutions before delays become losses, whether legitimacy survives changing guidance, whether lessons from one disruption alter the next budget cycle. A system with fewer visible assets may perform better than a richer one if its authority is clearer and its dependencies are understood. A richer system may still fail if every component waits for another component to move.

The path ahead follows that need for sharper tools. It begins by separating threat, hazard, and vulnerability so that strategic harm can be described without forcing every climate-linked disruption into the mould of hostile intent. It then tests the appeal of multiplier language and asks a stricter question about where amplification actually occurs inside institutions. From there the argument turns to the limits of deterrence as a model for feedback-driven risk, before settling on capability under climatic stress as a matter of logistics, staffing, information, legitimacy, and the fit between them.

What gives these concepts their weight is not elegance but consequence. Later discussions of infrastructure, budgets, migration, coordination failure, and Arctic governance will depend on whether the reader can see a hazard without mistaking it for an adversary, identify vulnerability without reducing it to moral weakness, and judge capability without taking official assurance at face value. The analytic burden grows from here. So does the political one. Once climate feedbacks are admitted into security thinking, the question is no longer whether governments can speak more forcefully about risk. It is whether they can speak more accurately while the queue is already forming.

That is the pressure point to keep in view before any narrower distinction is made. When a system fails under climatic stress, what exactly has happened: has nature simply struck, has a threat emerged, or has vulnerability already been built into the institutions meant to contain the damage? The answer will decide where responsibility is placed, which claims of preparedness deserve belief, and whether security analysis clarifies the problem or merely adds another line to the queue.

Threat, hazard, and vulnerability

In September 2017, Hurricane Maria struck Puerto Rico and tore through an electrical system already in poor condition. The storm was violent enough on its own terms: wind, rain, flooded roads, damaged substations, downed lines. Yet the strategic harm did not end with impact. Fuel distribution faltered. Backup generation became harder to sustain. Communications weakened. Hospitals and clinics struggled with interrupted power, damaged logistics, and uncertain resupply. The hazard was meteorological. The threat to public order, health, and state credibility emerged through the cascade that followed. The queue formed quickly.

That distinction matters because security language can mislead as easily as it can clarify. A hazard is a source of potential harm. A threat, in the stricter sense used here, is a condition that endangers assets, institutions, or social stability in ways that demand strategic attention. Political intent often intensifies threat, but it is not required for strategic harm. A storm has no motive. A thawing substrate makes no demands. Methane-related climate feedbacks do not calculate. Still, they can damage territory, strain budgets, weaken service delivery, and expose failures of public authority with consequences that feel unmistakably strategic. The absence of intent does not reduce the seriousness of the outcome. It changes the route by which it arrives.

The temptation, especially in public argument, is to compress these categories into one alarming word. Everything becomes a threat. Drought is a threat. Fire is a threat. Flood is a threat. Heat is a threat. Loose phrasing can create urgency, but it often blurs responsibility. If a storm is simply labelled a threat, one may overlook the institutional conditions that convert severe weather into prolonged social and political disruption. If climate change is described as a threat in the same manner as a hostile state or armed group, the language may borrow the prestige of security discourse while importing unhelpful assumptions about deterrence, retaliation, and control. Precision is not pedantry. It is a way of keeping the chain of causation visible.

Hazard is the easiest term to grasp because it begins with the physical world. A storm, a flood, a fire, a thaw, a period of dangerous heat: each presents a possibility of harm. One can map the hazard, track its intensity, estimate its reach, and compare it against known conditions. That work is indispensable. Yet hazards do not govern themselves. They meet roads, depots, pipelines, hospitals, housing, codes, budgets, and jurisdictions. They meet patterns of maintenance, settlement, and neglect. They meet the queue at the fuel depot. Once they do, the analysis must move beyond the event itself.

Threat enters when the consequences of hazard begin to endanger the continuity or credibility of systems that societies rely upon. A storm that damages roofs and trees is destructive. A storm that disables transport, interrupts fuel flows, impairs emergency response, and leaves a government unable to restore essential services in reasonable time produces a different order of problem. The distinction is not rhetorical inflation. It marks a shift from damage to strategic consequence. Critical services become less reliable. Trust erodes. Costs spread across sectors. Officials lose room to improvise. Markets adjust. Residents reconsider whether the state can still do what it claims. Threat, in this sense, is not simply severe hazard. It is hazard translated through vulnerability into institutional strain.

Vulnerability is the bridge concept, and it deserves more disciplined use than it often receives. It does not mean weakness in any vague or moralised sense. It refers to the characteristics that make a system more likely to suffer serious harm when exposed to hazard. A community dependent on a single access road is vulnerable in a way a better-connected community is not. A hospital reliant on backup generators without secure fuel resupply is vulnerable in a way a facility with diversified power arrangements is not. A housing stock built on unstable ground, maintained through deferred repairs, and insured under narrowing terms is vulnerable in a way a better-sited and better-funded stock is not. Vulnerability resides in the relationship between a system and the pressures acting upon it.

This is why climate security analysis goes wrong when it speaks only of hazard intensity. The same storm can strike two jurisdictions and produce markedly different outcomes because vulnerability differs. One region restores power, fuel, and transport with painful but manageable effort. Another faces prolonged outage, failed deliveries, overwhelmed hospitals, and a political crisis that outlasts the weather. The physical event matters. So does the institutional condition it encounters. Vulnerability is what explains why harm travels further in one place than another. It converts climate from background condition into a test of state performance.

A concrete comparison helps. Imagine severe wind damage in two coastal regions. In each, roads are blocked and power lines damaged. In one, fuel distribution resumes within a short window because depots are protected, priority routes are cleared according to rehearsed plans, mutual aid is interoperable, and emergency contracts can be activated without delay. In the other, tankers cannot reach damaged service stations, payment systems fail because communications remain down, local authorities lack authority to redirect stocks quickly, and generators at care facilities begin to run low. The storm damage may look similar from the air. The strategic meaning is not similar at all. The real divide lies in vulnerability.

Puerto Rico after Hurricane Maria made that point painfully visible. Public discussion often focused on the storm as the central cause, which it plainly was in immediate terms. But the prolonged severity of the crisis depended on more than meteorology. An electrical grid in poor condition, damaged transport routes, difficulties in fuel and equipment distribution, institutional strain, and weak administrative coordination all shaped the aftermath. Hazard initiated the event. Vulnerability governed the duration and depth of failure. The strategic implication is precise: reducing hazard exposure matters, but reducing vulnerability often determines whether a climate-linked shock becomes a temporary emergency or a prolonged erosion of public confidence.

This is especially important for methane-related climate feedbacks because feedback-driven warming often acts through systems whose fragility is easy to underestimate before they fail. A longer thaw season or more variable freeze-thaw cycle may not look dramatic in a briefing compared with the image of a fire front or flooded town. Yet those conditions can undermine roads, airstrips, storage sites, pipelines, and housing foundations over time. The hazard may appear incremental. The resulting threat emerges when a settlement loses dependable access, when a fuel corridor becomes uncertain, or when costly repairs crowd out other services. The damage may not arrive with the spectacle of a single disaster. It can still be strategic.

The line between hazard and threat becomes clearer when one asks a practical question: what exactly is being endangered? If the answer is immediate physical safety during an event, hazard may be the more useful term. If the answer includes continuity of service, legitimacy of governance, fiscal stability, or reliability of infrastructure networks, then threat has entered. Political intent remains absent. Strategic harm does not. That distinction allows climate-linked risks to be treated seriously without indulging in theatrical language. It also keeps attention on the institutions through which danger actually travels.

The phrase "security threat" often tempts writers and officials because it seems to promise seriousness. Yet seriousness purchased through conceptual looseness is a bad bargain. Call everything a threat and the category loses analytic value. Worse, it can flatten causal chains. A fuel shortage after a storm may be described as part of the same threat as the storm itself, obscuring the very decisions that made the shortage more or less likely: depot siting, reserve policies, transport redundancy, emergency procurement authority, maintenance of roads and ports, communications interoperability, staffing depth. Vague security language can therefore shift attention away from governance choices precisely when those choices are what matter most.

A second named example shows the same logic in colder terrain. In May 2020, a diesel tank near Norilsk in the Russian Arctic collapsed, releasing fuel into rivers and surrounding land. Public attention focused, understandably, on the spill and its environmental damage. Yet analytically the episode is useful for another reason. It suggests how a hazard linked to thawing ground conditions can become a strategic concern when critical infrastructure relies on environmental stability that no longer holds as assumed. The hazard is not simply "climate change" in the abstract. It is the altered condition of the ground. The threat arises when that altered condition compromises fuel storage, damages confidence in safety regimes, raises fiscal liabilities, and exposes how much of a remote system depends upon infrastructure whose foundations were treated as fixed.

This example also shows why vulnerability is not synonymous with poverty, remoteness, or low capacity alone. High-value industrial sites can be vulnerable if they depend on narrow assumptions about stable conditions, complex logistics, and limited repair windows. Conversely, some poorer or more remote communities may reduce vulnerability through local knowledge, diversified coping practices, or infrastructure choices better matched to place. Vulnerability is relational. It must be observed in the fit between system design and environmental stress, not inferred from broad labels. That matters because climate security analysis often swings between two lazy habits: assuming wealth guarantees safety, or assuming exposure guarantees collapse. Both can mislead.

Still, a limitation deserves emphasis. Not every hazard should be elevated into the register of strategic threat. Some events are severe but containable. Some disruptions remain local and temporary. Some institutions recover well precisely because vulnerability was reduced before the hazard struck. The Netherlands offers a useful counter-example in flood governance. Its exposure to water is obvious, yet decades of engineering, planning, and institutional development have altered vulnerability sufficiently that water hazard does not automatically translate into strategic crisis. That does not make the country invulnerable, and it does not settle every question about future risk. It does show that hazard alone cannot carry the full explanatory load. Threat depends on how the hazard meets the system.

That limitation strengthens, rather than weakens, the case for careful terminology. If not every hazard becomes a strategic threat, then analysts need a way to distinguish those that do from those that do not without relying on dramatic language or political convenience. Vulnerability provides that way. It asks where dependencies concentrate, where repair takes too long, where redundancy is absent, where authority is fragmented, where logistics are thin, where information arrives without decision rights, and where public trust has already been worn down by repeated mismatch between promise and performance. Once those features are visible, the line from physical change to institutional danger becomes much easier to trace.

Trust, often treated as an abstraction, becomes specific within this framework. It appears when residents obey evacuation orders because they believe fuel, shelter, and return planning are credible. It appears when clinics expect generators to be resupplied. It appears when local authorities share bad news early because they trust higher levels of government will not punish candour. It appears when insurers continue to write cover because standards are enforced and data are reliable. Each of these behaviours can fray when hazards repeatedly expose underlying vulnerability. A government that mistakes hazard management for vulnerability reduction may win a news cycle and lose its operating credibility.

Efficiency also looks different when hazard, threat, and vulnerability are kept distinct. A fuel system designed with minimal spare storage and tightly scheduled deliveries may appear efficient under ordinary conditions. After a storm or prolonged road disruption, that same design can prove highly vulnerable. Queues lengthen. Priority users compete with the public. Emergency vehicles draw from the same strained distribution network as everyone else. The hazard did not create the lean system. It revealed what the lean system could not absorb. Calling the outcome a "fuel threat" may not be wrong, but it is incomplete unless vulnerability is named as the bridge between event and consequence.

For climate feedbacks, especially those linked to methane's near-term warming pressure, this bridge is indispensable. Feedbacks can alter the pace and pattern of hazards without presenting themselves as a hostile actor. If warming affects permafrost stability, ice conditions, fire regimes, or hydrological extremes, the strategic question is not whether the atmosphere has intention. It is whether public systems are vulnerable enough that these changes produce failures in transport, energy, health care, communications, finance, or legitimacy. Hazard explains the pressure. Vulnerability explains the pathway. Threat describes the level of strategic consequence once the pathway begins to operate.

The practical value of this vocabulary is severe and simple. It disciplines diagnosis. Instead of asking whether climate change is "really" a threat, one asks which hazards are intensifying, where vulnerability is concentrated, and how institutional consequences will travel if current conditions persist. That approach avoids the false choice between minimising climate-linked danger as mere weather and inflating every disruption into a security emergency. It also exposes the quiet but decisive fact that many strategic harms are manufactured jointly by physical change and institutional design.

The queue is the image worth keeping. Not because it is dramatic, but because it is ordinary. A queue for fuel after a storm. A queue for repair crews after a thaw-damaged road closes. A queue for authorisations while agencies dispute responsibility. A queue for generators at a clinic whose backup plan assumed a delivery that never arrived. Hazards produce events. Vulnerability determines how long the queue becomes. Threat begins when the queue ceases to be an inconvenience and starts to measure the state's shrinking control over essentials.

That leaves a harder question than the language of security usually likes to ask. When officials describe a climate-linked danger as a threat, are they naming the hazard itself, or are they trying to avoid naming the vulnerability that made the harm strategically serious in the first place? Until that distinction is faced cleanly, governments will go on speaking forcefully about danger while leaving the queue exactly where it was.

Threat multipliers and their limits

In 2007, a report by CNA's Military Advisory Board helped popularise the phrase "threat multiplier" in security discussion of climate change. The phrase travelled well. It was brief, memorable, and strategically respectable. It suggested that climate did not replace older dangers but intensified them, aggravating instability where governance was already weak, resources contested, or social tension close to the surface. The appeal was obvious. A multiplying force sounds serious without demanding a claim that climate acts alone. It offers urgency with a measure of restraint. The shorthand stuck.

Yet useful phrases can become lazy ones. "Threat multiplier" often enters a discussion at the precise moment when analysis should become more exact. A drought contributes to migration pressure; migration pressure intersects with housing shortages, labour precarity, elite conflict, or border bottlenecks; political rhetoric then turns the resulting tension into something recognisable as security concern. That is not false. It is incomplete. The phrase tells us that pressure increased. It does not tell us where the pressure was carried, which institutions transmitted it, which decisions amplified it, or why one jurisdiction absorbed it while another spiralled. The wire is missing.

This matters because multiplier language performs two contradictory feats at once. It can elevate climate-linked risk into strategic view, which is often necessary in bureaucracies trained to treat environmental change as a secondary issue. It can also flatten causal chains into a vague drama of escalation. Once that happens, climate becomes the ominous backdrop to every difficult outcome. Governance choices blur. Institutional failures recede. Vulnerability is gestured at rather than traced. A phrase meant to sharpen attention begins to dull responsibility.

Take a familiar pattern after severe weather. A storm damages roads, power lines, and port access. Fuel distribution slows. Emergency generators start running through reserves. Food deliveries miss scheduled windows. Local authorities ask for assistance. Hospitals alter routine care. Public anger rises not because wind exists, but because ordinary systems no longer hold. Multiplier language might say the storm multiplied existing vulnerabilities. Fair enough. But which vulnerabilities? The condition of the fuel depots? The legal authority to redirect deliveries? The shortage of trained repair crews? The lack of spare transformers? The mutual aid agreement that assumed neighbouring regions would remain available? Unless those links are named, the phrase "multiplier" can disguise the real subject, which is institutional transmission.

The sharper question is therefore not whether climate multiplies threats in the abstract. The sharper question is: which institutions carry the multiplying effect, and when? That shift changes everything. It moves the analysis from broad atmospheric mood to operational sequence. A climatic pressure may multiply strain in health systems when heat and smoke raise admissions while power interruptions threaten cooling and communications. It may multiply strain in fiscal systems when repeated damage turns temporary repair into chronic expenditure. It may multiply strain in housing markets when displacement enters cities already short of affordable space. It may multiply strain in border administration when mixed flows meet slow processing and weak shelter capacity. The multiplier does not float above institutions. It passes through them.

That insistence on institutional location helps explain why the same physical hazard can produce such different political outcomes. A dry period in one region becomes a manageable stress on water allocation, insurance, and agriculture. In another, it contributes to rural income loss, urban migration, service deterioration, and sharpened conflict among local elites. To say climate multiplied threat in the second case is not wrong. It is merely the start of a serious explanation, not the end. The important work lies in specifying how a pressure reached a decisive threshold. Did it move through food prices, groundwater depletion, corruption in distribution, policing failures, displacement into informal settlements, or patronage networks that collapsed under scarcity? Each path implies different responsibilities and different forms of preparedness.

The Syrian drought of the late 2000s has often been cited in this debate, usually as part of an argument that climate stress contributed to the conditions from which later conflict emerged. The broad claim can be defended in cautious form: climatic stress affected livelihoods and migration patterns in ways that interacted with existing governance failures and social tensions. But the case also reveals the limit of multiplier language. It is too easy to move from "contributed to pressure" to "helped cause war" as though the institutional and political steps between them were mere detail. They are the substance. State choices about rural support, water management, repression, urban absorption, corruption, and violence shaped the path from environmental stress to political breakdown. Climate pressure mattered. So did the machinery through which it was filtered.

That is why multiplier language often performs poorly in public debate even when it began as a sensible corrective. It sounds precise while remaining permissive. Analysts can invoke it without committing themselves to a mechanism. Officials can use it to sound alert without specifying which ministry must change practice. Advocates can deploy it to dramatise urgency without distinguishing between domains where intervention is possible and domains where uncertainty remains high. The phrase starts to behave like an all-purpose solvent, dissolving fine distinctions into strategic atmosphere. Once everything is multiplied, little is truly measured.

A second example shows both the utility and the hazard of the phrase in a less conflict-centred setting. When Hurricane Maria struck Puerto Rico in September 2017, the storm's destructive force was undeniable. Yet the most enduring harm emerged through the prolonged failure of electricity, communications, transport, healthcare logistics, and fuel distribution. One could say the storm multiplied pre-existing weaknesses in infrastructure and governance. That is true. But it also leaves the impression that weakness was simply "there", awaiting multiplication, rather than composed of concrete decisions: deferred grid maintenance, fragile supply arrangements, difficult terrain, damaged roads, uneven administrative coordination, and the practical limits of restoring fuel and power under emergency conditions. If the analysis stops at multiplier, the lesson remains cloudy. If it names the institutions carrying the amplification, the lesson becomes operational.

Operational clarity matters because it changes what counts as preparedness. A government cannot prepare against a multiplication metaphor. It can prepare to protect fuel corridors, diversify power supply, pre-position repair assets, revise mutual aid assumptions, build fiscal contingency, decentralise certain decision rights, and strengthen communication between agencies. The more specific the causal chain, the more specific the corrective action. This does not mean certainty becomes easy. It means the field of responsibility becomes visible. Multiplication, in serious policy analysis, should refer not to a mysterious force of worsening but to a traceable increase in strain moving through identifiable systems.

The concept also struggles when it meets time. Multipliers are often presented as though they operate continuously and evenly, but institutional amplification is highly uneven. A climatic stress may remain manageable for months, then become acute when a budget cycle closes, an election delays action, a critical repair season is missed, or another hazard arrives before the first has been absorbed. Timing can turn a tolerable pressure into a strategic problem. A winter road with slightly reduced reliability may remain inconvenient until fuel planning, construction schedules, and household costs all hinge on the same shortening access window. The climate signal has not suddenly become morally worse. The institution's margin has vanished. What multiplies is not the hazard alone but the consequence of delay.

This is particularly important for methane-related feedback risk because the chapter's larger concern is not climate in general but the ways in which timing, non-linearity, and uncertainty press on state systems. Methane matters strategically when its warming influence sharpens near-term pressures on logistics, infrastructure, seasonal access, and fiscal planning. In that setting, multiplier language can obscure more than it reveals. It may suggest that climate simply intensifies existing problems in a broad sense, when the sharper question is where precisely the pressure lands. Does it hit a regional transport network already operating without spare capacity? A health system whose continuity depends on uninterrupted diesel supply? A municipal budget patched through repeated emergency appropriations? A housing market where new displacement immediately spills into hazard-prone land? These are institutional stories before they become strategic slogans.

Still, the phrase should not be discarded entirely. It captures something real. Climatic stress often does worsen outcomes by interacting with fragility rather than replacing it. The problem is not that the phrase is false.

The problem is that it is too easily treated as sufficient. Used carefully, it can mark the beginning of explanation: a warning that hazard and vulnerability are interacting. Used carelessly, it becomes a substitute for explanation. The discipline required is to let the phrase open the file, not close it.

A fair counter-example comes from settings where institutions absorb climatic pressure without obvious escalation. A heatwave may raise demand on health services and power systems, yet strong interagency coordination, reserve capacity, and clear public communication can prevent the event from becoming a wider strategic disturbance. In such a case, one could still say climate stress had multiplier potential, but the institutions carrying that potential did not fail. The amplification was interrupted. This matters because it shows that multiplier effects are contingent, not automatic. Pressure does not simply travel because it exists. It travels when systems permit it to travel.

That contingency returns us to governance. One reason multiplier language can be politically comfortable is that it appears to spread blame widely. Climate, after all, sits above everyone. The atmosphere becomes the stern background against which local failures unfold. But governments do not govern the atmosphere directly in the moments of crisis that matter here. They govern permits, depots, roads, staffing levels, budget reserves, legal authorities, data sharing, maintenance cycles, and emergency protocols. When analysts ask which institution carries the multiplying effect, they force the conversation back into the realm of choice. Was fuel distribution central? Was it school closure and childcare disruption? Was it municipal debt? Was it an overwhelmed reception system at the border? Once those questions are asked, multiplier language becomes a map prompt rather than a rhetorical flourish.

Budgets reveal the point sharply. Climatic pressure often multiplies fiscal strain not because heat, flood, or thaw possess economic cunning, but because public finance is organised around narrow categories and short horizons. Repeated small repairs crowd out preventive investment. Emergency appropriations mask chronic exposure. Insurance withdrawals shift liabilities onto public balance sheets. A local government defers maintenance to meet annual targets, then faces more expensive failure later. Here the multiplier is carried by the budget system itself. The pressure becomes strategic because financing rules and political incentives convert physical stress into cumulative weakness.

Infrastructure tells a similar story. A bridge, pipeline, road corridor, or fuel depot rarely fails in a single, isolated sense. Its condition interacts with inspection regimes, spare parts, contractor availability, communications, and access for repair crews. If multiple infrastructure systems depend on one corridor, then climatic disruption to that corridor multiplies through each dependent service. The institution carrying the effect may not be the transport ministry alone. It may be the procurement framework, the warehouse network, the local maintenance contractor base, or the fragmented authority between national owner and regional operator. The causal chain matters because infrastructure failure is often less about spectacular destruction than about hidden coupling.

Migration sharpens the same point. Climatic stress may influence movement by affecting livelihoods, safety, and local habitability. But whether mobility becomes politically explosive depends on receiving institutions: housing supply, documentation systems, school places, labour absorption, healthcare access, policing, and public narrative. To call climate a threat multiplier in migration politics is only a beginning. The real question is which institutions turn incoming pressure into manageable adaptation or into grievance and polarisation. Administrative capacity is not a side issue here. It is the site of multiplication.

Trust belongs in this account as well, but only when made concrete. Trust is not a civic virtue floating above institutions. It shows itself when residents comply with evacuation orders because fuel and shelter plans have worked before, when hospitals believe emergency supply requests will be met, when local officials share bad news early because higher authorities do not punish candour, when migrants enter formal systems because processing is credible, and when treasuries fund preparedness because prior warnings proved institutionally useful rather than politically embarrassing. Climatic pressure can multiply distrust, but only through repeated experiences of institutional mismatch. The breach occurs when promise and performance diverge often enough to teach people that contingency plans are merely printed confidence.

So the phrase "threat multiplier" should be handled like a warning label, not a diagnosis. It alerts us to interaction. It should then compel a more demanding inquiry. Which institution carries the amplified strain? At what point in the sequence does manageable stress become strategic harm? What visible behaviour marks that shift: queues at fuel depots, school closures, emergency procurement waivers, insurer retreat, repeated power interruptions, municipal borrowing, overcrowded shelters, delayed repairs, contradictory public guidance? Once the analysis turns to those particulars, climate-security argument becomes less dramatic and more useful.

The decisive question is no longer whether climate multiplies threats. It plainly can. The decisive question is whether governments can identify the exact channels of multiplication before those channels harden into routine crisis. If they cannot, the phrase will continue to flatter strategic discourse while leaving the real carriers of danger untouched: the budget line, the road corridor, the depot, the processing office, the maintenance backlog, the quiet institutional seam where pressure ceases to be weather and becomes weakness.

Deterrence does not apply cleanly

In October 1962, during the Cuban Missile Crisis, the United States and the Soviet Union confronted one another through a logic both sides understood with terrifying clarity. Capabilities were counted. Signals were sent. Intent was read and misread. The danger lay in what one adversary might choose to do if the other failed to persuade, threaten, or retreat. However unstable the moment, the grammar was recognisable. Deterrence belonged there because the crisis turned on calculation: who could impose unacceptable costs, who believed the threat, and who would blink. The switch was visible.

That grammar has shaped much of modern strategic thought. It has given policymakers and analysts a language for escalation, coercion, credibility, signalling, red lines, retaliation, and resolve. Those concepts are not obsolete. They remain central wherever organised actors seek to alter one another's behaviour through threat of punishment or denial. Yet climate feedback risk does not enter political life through that route. Methane does not read communiques. A thawing substrate cannot be warned off. A heatwave does not recalculate after a show of force. Uncertainty does not negotiate. The switch does not respond.

This sounds obvious when stated so baldly, but the pull of deterrence thinking remains strong because security institutions inherit habits as much as doctrines. When a problem is labelled strategic, the mind reaches quickly for the familiar tools of strategic analysis. Who is the adversary? What is the signal? What would count as resolve? Where is the red line? How can one deter the next escalation? Such questions are useful when dealing with hostile states, armed groups, or coercive economic behaviour. Applied to feedback-driven climate risks, they begin to distort. They privilege confrontation over preparedness, drama over maintenance, declaratory posture over civil capability. The result is not merely a category mistake. It can become a governing mistake.

The distinction turns on intent. Deterrence presumes an opponent capable of perceiving threatened costs and altering behaviour accordingly. Even deterrence by denial, which seeks to make aggression futile rather than punish it afterward, still assumes an actor deciding whether the effort is worth the likely result. Climate feedbacks do not operate in that moral and strategic space. They act through physical processes, threshold effects, compounding stresses, and institutional vulnerabilities. Harm emerges not because an enemy chooses to impose it, but because systems are exposed, sensitive, and poorly positioned to absorb pressure. Strategic consequence remains real. Strategic reciprocity does not.

That is why retaliation fits so badly. Retaliation is meaningful when directed at an actor whose future conduct may be shaped by anticipated punishment. If a state suffers infrastructure damage because unstable ground, extreme heat, or compounding hazards undermine systems on which public order depends, there is no adversary to punish into restraint. One may regulate emitters, negotiate internationally, litigate liability, or sanction deception in environmental reporting. Those are distinct political and legal acts, sometimes justified, sometimes not. They are not retaliation in the classical strategic sense against the hazard itself. Confusing the two blurs the difference between accountability for human choices and the management of physical instability already in train.

A named case helps sharpen the point. When Hurricane Sandy struck the eastern United States in October 2012, the immediate issue for New York and neighbouring regions was not signalling resolve to an adversary. It was restoring transport, electricity, fuel distribution, communications, and public confidence under severe stress. Petrol queues became one of the most memorable images of the aftermath in New York and New Jersey. Those queues mattered because they revealed how quickly an urban system could become strategically fragile once electricity, transport access, fuel logistics, and administrative coordination strained together. No doctrine of deterrence could have prevented those queues. The relevant questions lay elsewhere: where were the bottlenecks, what reserves existed, which authorities could prioritise deliveries, how dependent were critical services on uninterrupted fuel flow, and how rapidly could institutions adjust under pressure? The language of coercion offered little help.

This is not an argument for stripping climate-related security analysis of seriousness. Quite the opposite. It is an argument for protecting seriousness from theatrical borrowing. Military vocabularies exert prestige because they suggest discipline, urgency, and consequence. Sometimes that borrowing is harmless. Often it is not. If a government speaks of climate resilience as though it were a theatre of deterrence, institutional attention may drift towards visible posture and away from the quiet work that actually determines outcomes: drainage standards, backup power, staffing depth, mutual aid realism, warehouse locations, interoperable communications, clear legal authority under emergency conditions, and the maintenance of public trust through candid revision. The slogan hardens. The depot stays empty.

The danger deepens when policymakers import not only the language but the emotional habits of deterrence. Deterrence thinking prizes credibility, consistency, and the avoidance of signals that might invite challenge. Those instincts make sense in adversarial settings where hesitation can indeed be exploited. Under feedback-driven risk, the same instincts can become liabilities. Officials may cling to outdated assurances because admitting uncertainty feels like weakness. They may avoid revising guidance because visible adjustment resembles retreat. They may prefer strong declaratory statements to openly conditional planning because ambiguity seems politically costly. Yet uncertainty in climate-linked systems is not a bluff to be called. It is a condition to be governed. A state that cannot revise without embarrassment will often revise too late.

The COVID-19 pandemic, though not a climate event, offered a vivid demonstration of this institutional trap. Governments that treated changing guidance as a reputational threat sometimes delayed adjustment, defended stale assumptions, or communicated with a confidence the evidence did not warrant. The issue was not deterrence in any formal sense, but the pathology was familiar: a fear that visible adaptation would look like weakness. The lesson transfers. Under deep uncertainty, credibility does not rest on pretending not to revise. It rests on establishing in advance that revision is part of competence. That is a profoundly un-deterrent idea. It values adaptation over consistency for its own sake.

A second example comes from northern infrastructure. In May 2020, the diesel spill near Norilsk entered international discussion because the collapse of a storage tank raised questions about thawing ground conditions beneath critical facilities. However one apportions the causes, the strategic meaning of the event lies in what it exposed: dependence on infrastructure whose foundations could no longer be treated as passive constants. No act of resolve could deter unstable ground from becoming more unstable. No retaliatory posture could restore confidence in a maintenance regime once the failure had occurred. The strategic task lay in something plainer and harder: inspecting, reinforcing, redesigning, funding, monitoring, and deciding earlier under uncertainty. When the hazard is non-sentient but the consequences are institutional, preparedness replaces coercion as the decisive domain of statecraft.

This is where some analysts attempt a rescue operation by broadening deterrence until it means almost any effort to reduce harm. One sometimes hears that adaptation can be understood as a kind of deterrence by denial: harden enough infrastructure and the threat cannot achieve its effect. The analogy has surface appeal. It preserves familiar terminology while shifting emphasis from punishment to prevention. Yet it still misleads. Denial in military strategy works because an opponent can observe obstacles and reassess the wisdom of attacking. A flood, wildfire, thaw, or compound heat-and-power event does not make that assessment. Better to speak plainly. Some measures reduce exposure. Others reduce sensitivity. Others increase response capacity. That language may sound less dramatic, but it maps more accurately onto what institutions are actually doing.

Plain language also clarifies what "strategic" should mean here. Strategic does not require an enemy commander. It refers to harms or decisions that affect the continuity of essential functions, the credibility of public authority, the durability of infrastructure, the allocation of scarce resources, and the ability of a state to preserve room for manoeuvre under pressure. A fuel distribution system becomes strategic when its failure disables emergency services, transport, food supply, and hospital continuity at once. An Arctic runway becomes strategic when its degradation compromises evacuation, medical access, and sovereignty expressed through service provision. A housing system becomes strategic when recurrent damage and insurance retreat alter municipal finance, migration, and public trust. The strategic quality lies in consequence and interdependence, not in hostile intent.

That matters because once "strategic" is detached from the narrow image of adversarial conflict, the hierarchy of relevant questions changes. Instead of asking what signal will dissuade the next challenge, analysts ask which dependencies are least forgiving, which assets are hardest to replace, which thresholds would trigger cascading failure, which legal authorities become binding constraints under emergency conditions, which information needs to travel fastest, and which capacities are falsely assumed to exist because they are written into a plan. These are not lesser questions than those of classical strategy. They are the strategic questions appropriate to a different class of danger.

Still, the contrast should not be overstated. There are domains where deterrence and climate risk genuinely intersect. States may compete over Arctic access, exploit disaster conditions for political advantage, weaponise migration narratives after climate-linked displacement, or engage in sabotage against infrastructure already stressed by environmental change. In such cases adversarial behaviour and physical hazard coexist. Military frameworks then have their place, but only with care. The relevant point is not that deterrence disappears from the wider security landscape once climate feedbacks are considered. It is that deterrence does not govern the feedback risk itself. Analysts must keep separate what belongs to hostile intent from what belongs to institutional preparedness under environmental uncertainty. Otherwise the two domains blur into one another and neither is well served.

This is the limitation worth keeping. Not every use of military analogy is crude. Armed forces possess planning disciplines, logistical expertise, and scenario methods that can contribute valuably to national resilience. Defence institutions often understand redundancy, degraded operations, and the practical significance of supply lines more clearly than civilian bureaucracies do. The mistake lies not in learning from such traditions but in letting their coercive concepts dominate problems where coercion has no target. A logistics table may travel well. A retaliatory logic does not.

The practical cost of confusion appears in budgets and institutions. A government captivated by strategic rhetoric may invest in visible symbols of seriousness while neglecting mundane capacities that actually determine continuity under climate stress. It may overvalue central command and undervalue local authority. It may underfund maintenance because maintenance lacks drama. It may tolerate brittle just-in-time systems because their peacetime efficiency flatters managerial culture. It may frame public communication as a contest of confidence rather than a process of conditional updating. Each choice looks small. Together they create a state better at sounding prepared than at remaining governable when conditions deteriorate.

Trust reveals the difference. In a deterrent relationship, credibility often rests on convincing an opponent that threats will be carried out. Under feedback-driven risk, credibility rests on a different set of visible behaviours:

warnings issued early, guidance revised openly when conditions change, backup systems tested rather than merely declared, mutual aid agreements exercised under realistic assumptions, infrastructure repaired before failure becomes routine, local observations treated as information rather than nuisance, and emergency powers tied to clear legal authority. These acts do not signal resolve to nature. They signal competence to the public. That is the relevant audience.

Efficiency also changes its meaning once deterrence falls away. In adversarial settings, visible capability may deter precisely by being costly and unmistakable. Under climate stress, visible capability may matter less than the spare part in storage, the technician retained on payroll, the culvert replaced before the wet season, the fuel reserve rotated properly, the authority to close a road before it fails, or the procurement rule that allows rapid substitution when a supplier collapses. None of this is glamorous. Much of it looks inefficient on paper. Yet strategic weakness often begins where such capacities have been shaved away in the name of tidy budgets and managerial confidence.

The deeper issue is one of institutional temperament. Deterrence rewards firmness, clarity, and the credible threat of consequence. Governing uncertainty rewards something more difficult: firmness about purpose, flexibility about means, honesty about what is not known, and willingness to revise before events force humiliation. One posture communicates strength through fixity. The other communicates strength through controlled adaptation. The two are not morally opposed, but they are not interchangeable. A state that mistakes one for the other may discover that its strongest voice has been reserved for the wrong audience.

That leaves a more exact definition of strategic life under climate feedback risk. Strategic judgment here concerns where to preserve slack, when to act without full confidence, which dependencies deserve protection because their failure multiplies harms elsewhere, and how to maintain legitimacy while public guidance changes under pressure. None of that can be settled by threats issued into the atmosphere. It must be settled in depots, standards, contracts, maintenance cycles, transport plans, reserve funds, and the political willingness to explain why visible caution is not weakness but statecraft of the correct kind.

The question is not whether governments can speak about climate in the language of security. They plainly can. The question is whether they can resist the seduction of a security language designed for opponents who can hear them. If the adversary is uncertainty, what proves strength: a louder warning, or a state arranged so that when the next shock arrives, the switch that matters is not fear but function?

Measuring capability under climatic stress

In August 2023, the evacuation of Yellowknife in Canada's Northwest Territories revealed a familiar and uncomfortable truth about state performance under pressure. The hazard was wildfire. The visible questions seemed immediate enough: who could leave, by which route, with what assistance, and in what order. Yet the deeper test lay elsewhere. It lay in whether transport, health services, communications, fuel supply, shelter arrangements, and public messaging could be aligned under narrowing time and imperfect information. Capability showed itself not as a slogan, but as a sequence. The hand-off mattered.

That sequence is the right place to begin because capability is so often described too loosely. Governments say they are prepared. Agencies say they have capacity. Emergency plans cite personnel, vehicles, agreements, stockpiles, and command structures. Much of that can be true and still insufficient. Climatic stress does not merely ask whether resources exist in the abstract. It asks whether resources can be moved, authorised, shared, trusted, and sustained while several systems are under pressure at once. A warehouse on paper is not a delivered generator. A mutual aid agreement in a binder is not an available crew when neighbouring jurisdictions are facing the same crisis. The hand-off decides.

Capability, then, should be treated as a function of four interlocking elements: logistics, staffing, information, and legitimacy. Each sounds familiar. Each is commonly misunderstood. Logistics is often reduced to transport alone, as though roads and aircraft were the whole story. Staffing is often reduced to headcount, as though any trained person could step into any role under pressure. Information is often reduced to data availability, as though knowing more automatically produced timely action. Legitimacy is often reduced to public sentiment, as though trust were merely a mood. Under climatic stress, each of these abstractions becomes concrete. They become routes, rosters, dashboards, decision rights, queues, and compliance.

Logistics comes first because it is the hidden architecture of capability. A state can possess excellent plans, advanced modelling, and sincere political intent and still fail because it cannot move people, fuel, food, spare parts, and repair crews where they are needed quickly enough. Climatic stress exposes logistical truth with unusual cruelty. A washed-out road, a smoke-choked flight path, a damaged port, a softening runway, or a shortened winter access window can reduce practical capability long before formal authority recognises the change. This is why preparedness cannot be measured only by what exists in inventory. It must be measured by what can be reached, sustained, and replaced once disruption begins.

Hurricane Maria in Puerto Rico in September 2017 remains one of the clearest modern examples. The storm caused immense damage. That fact is straightforward. The strategic lesson lies in what followed. Power restoration faltered, roads were blocked, fuel distribution was disrupted, communications failed, and hospitals struggled to maintain continuity. The decisive problem was not merely the number of generators or fuel reserves somewhere within the territory. It was whether fuel could be transported, prioritised, and delivered to the places whose operation held other services together. A hospital without diesel is not a hospital for long. Capability in such circumstances is logistical before it is rhetorical.

The same logic applies in slower, less spectacular settings. A northern jurisdiction that depends on winter roads for bulk deliveries may appear adequately supplied in annual budget documents. Yet if the operating season narrows or becomes unreliable, the capability picture changes sharply. Stock must be moved earlier, in larger volumes, at greater cost, or by more expensive alternatives. Delayed shipments alter maintenance schedules and household costs. Public agencies must choose what to prioritise. The relevant measure is not simply whether a route exists, but whether it remains dependable enough for the systems attached to it. Climatic stress often turns logistics from background assumption into front-line determinant.

Staffing is the next element, and it is equally easy to misread. Headcount matters, but only crudely. Capability depends less on how many people are employed than on whether the right people are available, authorised, rested, retained, and able to work together under altered conditions. A jurisdiction may appear well staffed in ordinary time and still prove dangerously thin under prolonged fire smoke, overlapping evacuations, infrastructure failure, or repeated emergency activations. Specialised roles matter disproportionately here: line crews, air traffic staff, dispatchers, public health coordinators, fuel technicians, emergency physicians, procurement officers, and local leaders who know how informal realities diverge from formal plans. Lose a few of the wrong people and the system slows far more than aggregate staffing numbers suggest.

This is one reason mutual aid is so often overpraised. Mutual aid agreements can be valuable. They can bring in crews, equipment, shelter support, and technical expertise that a single jurisdiction could never maintain alone. But their success depends on an unstated assumption: someone nearby remains able to help. Under regional simultaneous events, that assumption fails. Wildfire seasons offer repeated examples. When smoke, heat, and fire stretch across multiple regions at once, neighbouring jurisdictions may be unwilling or unable to release personnel and assets they would normally share. The agreement remains valid on paper. The practical capability has vanished. This is not a legal defect. It is a planning defect born of optimistic assumptions about geographic staggering.

The problem becomes more acute when staffing is tied to narrow procurement and budget rules. Many public systems have shed slack over decades in pursuit of visible efficiency. Vacancy controls, temporary contracts, reduced maintenance teams, consolidated control rooms, and just-in-time contracting may all appear rational in calm periods. Under climatic stress they can prove ruinous. A vacant technical post can delay a repair contract. A fatigued dispatcher can slow a whole emergency network. A contractor pool trimmed too aggressively can leave no one available when several assets fail at once. Efficiency, in this context, must be judged by continuity under strain, not by peacetime neatness. The roster tells a story.

Information is the third element, and here a different confusion prevails. Institutions often behave as though better data naturally produce better decisions. Sometimes they do. Often they do not. Monitoring systems may detect rising river levels, thawing ground, smoke trajectories, infrastructure anomalies, or service interruptions with admirable precision. Yet capability depends on whether information is timely, shareable, credible, and attached to authority. A dashboard no one can act upon is not capability. A forecast locked in a separate agency, a warning delayed by clearance procedures, or a local observation dismissed because it does not fit central reporting formats can all reduce practical preparedness even where the technical information base is strong.

The 2021 flooding in British Columbia illustrated how information and action can diverge under compound strain. The province faced severe disruption to highways, rail lines, supply routes, and communities after intense rainfall and landslides. Much was known about weather hazards and vulnerable corridors in general terms. The harder problem lay in how quickly information about damage, route status, supply consequences, and emergency priorities could be turned into coordinated action across sectors and levels of government. Information does not become capability merely by existing. It becomes capability when institutions know who can decide, who must be informed, and what pre-agreed action follows a changing signal.

This leads directly to decision rights, which are often neglected when capacity is discussed. A well-funded organisation with skilled staff and good data can still fail if no one has clear authority to trigger action under uncertainty. Who can close a road early? Who can reroute fuel? Who can override routine procurement rules during cascading service disruption? Who can order pre-emptive evacuation transport for people whose needs are not captured by generic plans? Decision rights may sound procedural, but under climatic stress they become operational. Delay is often less a shortage of knowledge than a shortage of authorised action. A capable system makes such rights visible before the emergency, not during it.

Interoperability belongs here as well. Public agencies frequently measure their own readiness vertically: the transport ministry knows its assets, the health system knows its beds, the electricity operator knows its substations, the local authority knows its shelters. Climatic stress punishes vertical confidence. It exposes whether agencies can actually work across boundaries of jurisdiction, software, terminology, command culture, and legal mandate. A hospital may have a sound backup plan that assumes fuel deliveries the transport authority cannot guarantee. A municipality may open shelters without live information about power restoration timelines. A provincial emergency office may issue public guidance that conflicts with local operational knowledge. Capability exists in the fit between systems, not in their isolated competence.

Legitimacy is the fourth element, and perhaps the most underestimated. It is tempting to treat legitimacy as a soft variable, interesting but secondary to hard assets. That is a mistake. Under climatic stress, legitimacy becomes practical because it shapes compliance, candour, and cooperation. People evacuate when they believe the route, shelter, and return arrangements are credible. Local officials speak plainly when they trust that higher authorities will support rather than punish early warning. Contractors mobilise when payment and command structures are reliable. Communities share information when they believe it will be used responsibly. Legitimacy, then, is visible behaviour. It is not measured by slogans about trust. It is measured by whether the system functions more smoothly because people believe participation is worthwhile.

COVID-19 made this painfully plain in many countries, though the hazard there was epidemiological rather than climatic. Guidance changed as evidence evolved, and the ability of institutions to maintain public cooperation depended heavily on whether revisions were explained as competence or experienced as confusion. The parallel matters. Climate-related emergencies also require changing guidance under uncertainty. A jurisdiction that treats updates as embarrassment will often communicate too late or too rigidly. A jurisdiction that treats revision as normal practice stands a better chance of preserving legitimacy while conditions change. Capability therefore includes the cultural ability to update without collapse in confidence.

Still, capability cannot be reduced to institutional grace under pressure. Resources matter, and any argument that downplays them becomes sentimental. Wealthier states and regions can buy lift capacity, harden infrastructure, maintain reserves, and retain specialised staff more easily than poorer ones. A small municipality cannot conjure a deep contractor base or redundant transport network simply by improving coordination. This is a real limitation, and it should be faced honestly. Capability is not a moral quality awarded to prudent administrators. It is materially conditioned. Yet material strength alone does not settle the matter. Rich systems can be brittle. Poorer systems can sometimes perform better than expected because they preserve local knowledge, clear authority, and habits of adaptation that richer but more fragmented systems have allowed to decay.

A useful counter-example lies in the Netherlands, where flood management has long demonstrated how capability can be built through institutional continuity rather than mere episodic spending. The country benefits from wealth and technical expertise, certainly. But its performance also depends on clearer lines between monitoring, engineering, planning, and long-term political commitment than many jurisdictions sustain. The lesson is not that all climatic risks can be managed as Dutch water risk is managed. They cannot. The lesson is that capability grows when information, authority, maintenance, and legitimacy are arranged into a durable pattern rather than activated only after crisis. Capacity is organised memory.

This raises the question of indicators. Preparedness claims deserve scrutiny, and scrutiny requires signs more informative than public confidence alone. One should ask whether logistics have genuine redundancy or only optimistic routing plans; whether critical staff roles are filled and retained or merely budgeted for; whether agencies share interoperable data or simply issue parallel reports; whether decision rights are clear under uncertain conditions; whether exercises test simultaneous regional strain rather than neat local scenarios; whether after-action reviews change procurement, standards, or staffing; whether mutual aid assumptions include the possibility that neighbours are also overwhelmed; whether local observations can alter central plans. These are not glamorous indicators. They are better than glamorous ones.

The attraction of simplistic metrics is understandable. Governments like numbers they can publish. Agencies like readiness scores, completed exercises, stockpile counts, and expenditure totals. Such measures are not useless. They become misleading when treated as proxies for capability rather than fragments of it. A fully stocked depot means little if a washed-out road blocks access. A completed interagency exercise means little if the scenario assumed only one jurisdiction under pressure. A funded post means little if recruitment repeatedly fails. A public information campaign means little if communities no longer believe official timings. Capability has to be measured in relation, not merely in volume.

That relational quality is what climatic stress reveals most harshly. Heat, smoke, flood, thaw, fire, and transport disruption do not respect administrative boundaries. They test the state as a linked system. A government may discover that its apparent strengths are arranged in the wrong order: excellent forecasting but poor decision rights, ample equipment but thin staffing, detailed plans but weak fuel logistics, strong central authority but poor local legitimacy. In such conditions the issue is not whether capacity exists somewhere in the apparatus. It is whether the parts can reach one another in time. The hand-off fails, and the public experiences failure as one undifferentiated fact.

Measuring capability under climatic stress therefore requires a cooler and more suspicious eye than public discourse often encourages. It demands that claims of readiness be traced into routes, contracts, rosters, authorities, repair cycles, mutual aid realism, communications compatibility, and visible public compliance. It also demands acceptance of a harder truth: some capabilities only reveal themselves when several assumptions break together. The question is not whether a system looks impressive when each agency reports separately. It is whether, when the smoke thickens, the road closes, the backup fuel runs low, and the neighbouring jurisdiction stops answering because it is in trouble too, the hand-off still holds.

Chapter 3

Methane Systems and Monitoring Limits

What counts as knowledge when the quantity in dispute depends partly on how it is measured, partly on where one looks, and partly on which institution is doing the looking? That question sits at the centre of methane politics more firmly than public argument usually admits. Policymakers often ask for a simple figure, a stable baseline, a settled account of source and scale before they commit money, revise standards, or defend cooperation. Yet the methane problem does not arrive in that obedient form. It comes instead with competing inventories, partial observations, uneven monitoring coverage, modelling choices, and a public atmosphere quick to confuse disagreement over method with proof of deception. The reading changes.

This is not a trivial technical complication. It is a strategic condition. States make expensive decisions on the basis of what they think they know about assets, exposure, and timing. If methane systems are measured differently by inventories, field campaigns, atmospheric sampling, and remote sensing, then the issue is not merely scientific housekeeping. It becomes a question of credibility, budgeting, and institutional trust. A government deciding whether to harden infrastructure, expand monitoring, or negotiate internationally is not choosing in a world of pristine knowledge. It is choosing under contested visibility. The map is drawn while the route is being argued over.

That tension sharpens in the Arctic, where distance lends false simplicity to public imagination. High-latitude change is often spoken of as though it were geographically remote and politically self-explanatory: ice retreats, ground thaws, strategic interest rises. The reality is harder. What matters is not only that warming at high latitudes draws attention from distant capitals, but that it does so through the practical instability of ground conditions, maintenance cycles, transport assumptions, and service provision. Permafrost, in this setting, is not scenery. It is substrate. Once the substrate becomes less reliable, every confident statement about a facility, a runway, a storage site, or a remote settlement carries a hidden qualification.

The central tension therefore runs deeper than the familiar contrast between science and politics. Monitoring systems promise more visibility, but visibility alone does not settle action. Better sensors, more satellites, denser observation networks, and improved modelling can sharpen trends, detect anomalies, and narrow attribution ranges. They cannot decide which ministry acts, what threshold triggers spending, which level of government bears liability, or how uncertainty should be explained without inviting either panic or complacency. Information that floats free of decision pathways can become corrosive. It frustrates officials, feeds suspicion, and leaves publics with the impression that warning has increased while competence has not.

This matters because the argument over methane is often framed as though the decisive battle were simply over facts. Facts matter. So do methods, mandates, and timing. A baseline is never just a number when that number will shape international claims, domestic budgets, regulatory credibility, and the perceived seriousness of risk. A monitoring alert is never just an alert when no one has agreed in advance what follows from it. Public trust is not maintained by promising certainty where none exists; it is maintained when institutions show that uncertain knowledge still has a disciplined place inside action. That discipline is visible in thresholds, reporting lines, maintenance decisions, procurement rules, and public communication that can change without collapsing into contradiction.

The stakes reach further than the atmosphere. When governments cannot explain why different measurement systems yield different pictures, disagreement is easily recast as bad faith. When agencies collect data but cannot translate it into authority, monitoring begins to look ornamental. When uncertainty is communicated badly, citizens hear either evasion or alarmism. The strategic cost is not only confusion. It is degraded compliance during disruption, weaker support for preventive spending, and greater vulnerability to politicised narratives that treat complexity as proof of incompetence. A state does not need perfect knowledge to govern well, but it does need a credible account of what is known, what is contested, and who is responsible for acting under both conditions.

The path ahead follows that problem from quantity to consequence. It begins with the difficulty of establishing methane baselines when sources, sinks, and methods do not align cleanly, then moves northward into the specific relevance of Arctic amplification and unreliable ground conditions for built assets and remote planning. From there the focus shifts to satellites, sensors, and the institutional machinery that turns observation into usable governance, before arriving at the public question that shadows all the others: how uncertainty can be spoken about without paralysis, distrust, or the pretence that ambiguity has been abolished.

What gives these themes their coherence is not scientific novelty alone, but a recurring administrative dilemma. States want enough certainty to justify commitment, yet many of the most consequential decisions must be made before that certainty arrives. They want monitoring systems that improve foresight, yet foresight without delegated authority often heightens blame instead of reducing harm. They want public trust, yet trust is damaged both by false confidence and by communication so cautious that it sounds like withdrawal. The reading changes again: not because nature is fickle in some poetic sense, but because institutions must decide what counts as adequate knowledge before they can decide what counts as adequate preparation.

That is the pressure point to keep in view before the finer distinctions begin. When officials ask how much methane matters, they are rarely asking only about molecules. They are asking what can be claimed with confidence, what can be budgeted without ridicule, what can be monitored without creating another layer of delay, and what can be said in public without losing either credibility or control. The hard question is not whether better measurement is desirable. It is what kind of state emerges when the reading changes but the decision pathway does not.

Sources, sinks, and contested baselines

In November 2022, the United Nations Environment Programme released its International Methane Emissions Observatory report into a policy climate already crowded with confident claims about where methane comes from and how much of it is escaping. The institution was named. The date is easy to verify. The subject looked numerical, even tidy. Yet the deeper story was not tidy at all. The same atmosphere could be described through national inventories, company reporting, aircraft campaigns, satellite observations, field sensors, and modelling exercises that did not always line up. The number moved.

That movement is not a failure of seriousness. It is part of the subject. Methane enters public life through several major source categories, each measured with different levels of ease and dispute. Fossil fuel production and transport matter because leaks, venting, and incomplete combustion can release methane across wells, pipelines, compressor stations, storage sites, and distribution systems. Agriculture matters because livestock and manure management generate emissions through biological processes spread across large areas and varied husbandry systems. Waste matters because landfills and wastewater can produce methane under conditions that change with temperature, moisture, and management practice. Wetlands matter because they are major natural sources whose emissions vary by season, hydrology, temperature, and location. The atmosphere receives methane from all of these routes. Policy, however, does not encounter them on equal terms.

This is why baselines become contested so quickly. A baseline sounds like a simple starting point, a settled quantity against which change can be measured. In practice it is an arrangement of methods, assumptions, time windows, and categories. A national inventory may estimate emissions by combining activity data with standard emission factors. An aircraft campaign may sample plumes over a producing region and reveal emissions that appear larger or differently distributed than the inventory suggested. A satellite may detect strong point sources under favourable conditions but miss smaller diffuse releases or struggle with cloud cover, background concentration, or spatial resolution. A model may reconcile several streams of evidence and still depend on assumptions about transport, chemistry, or source attribution. The baseline is therefore never just a number. It is a way of seeing.

That matters because "how much" is often inseparable from "how known". When officials ask how large a methane problem is, they are usually asking two questions at once. They want to know the scale of emissions or uptake, certainly. They also want to know how confident they can be in the estimate, how frequently it can be updated, whether it is comparable with other jurisdictions, whether it will withstand scrutiny, and whether acting on it now will later look prudent or premature. The measurement problem is not an awkward prelude to policy. It is already inside policy. A baseline that appears stable enough for a ministerial statement may be too coarse for site-level enforcement. A plume detected by remote sensing may be vivid enough for headlines and too partial for national accounting. The number moves because the methods move.

The distinction between sources and sinks deepens the challenge. Sources add methane to the atmosphere. Sinks remove it, chiefly through atmospheric chemical reactions, while soils can also play a role under certain conditions. For public argument, sources tend to dominate because they are easier to attach to visible activities, industries, or landscapes. Sinks are less politically legible. They involve processes that are harder to personify and harder to fold into a simple blame narrative. Yet they matter because the atmospheric outcome depends on both addition and removal. A baseline that talks as though emissions alone settle the issue can create a false impression of certainty. It captures something real and still misses part of the system.

Show the problem in a familiar policy setting and the stakes become clearer. A government wants to set a methane reduction target for the oil and gas sector. It turns to reported data from operators, national inventory methods, and independent monitoring. The reported data suggest one pattern. Independent observations suggest another. Neither side need be fabricating. Operators may be counting according to established reporting rules that rely on standard assumptions. Independent observers may be detecting intermittent high-emitting events that the reporting framework does not fully capture or time-average in the same way. The concept to name here is methodological mismatch. Different tools answer different questions. The implication is severe: a dispute that looks political may be partly epistemic, but it will still have political consequences once budgets, compliance obligations, and public credibility are at stake.

A named example from the United States illustrates this well. In 2018, a study led by researchers associated with the Environmental Defense Fund and several academic institutions argued that methane emissions from the United States oil and gas supply chain were significantly higher than estimates derived from the Environmental Protection Agency inventory methods then in use. The finding drew attention not only because of the result, but because of what produced the difference. The study used extensive direct measurements across segments of the supply chain, while the inventory system relied heavily on bottom-up estimation methods and standard factors. The disagreement did not automatically prove bad faith by either side. It exposed how different measurement architectures can produce different pictures of the same sector. That exposure matters because governments build regulations, inspection priorities, and diplomatic claims on whichever picture they decide to trust.

A second example sits outside fossil fuel systems and shows why the dispute cannot be reduced to industrial reporting alone. Wetlands are among the world's major natural methane sources, yet they are notoriously difficult to measure consistently across space and time. In Arctic and sub-Arctic regions, thaw, hydrology, vegetation change, and temperature interact in ways that complicate attribution and trend detection. Field campaigns may identify strong local emissions. Satellite observations may suggest broader patterns. Models may interpolate beyond sparse ground measurements. None of this means the science is empty or arbitrary. It means natural systems do not conveniently organise themselves around the reporting needs of ministries. The baseline becomes sensitive not simply because politics interferes with science, but because the object being measured is variable, seasonal, and distributed unevenly across difficult terrain.

This is where arguments about contested baselines can turn dishonest if handled badly. One temptation is to treat disagreement between methods as proof that no one knows anything. Another is to treat one favoured method as the uniquely true picture and all others as evasions. Both responses are analytically poor. Inventories are useful because they provide structured, repeatable accounting frameworks that states can use for reporting and planning. Measurement campaigns are useful because they can reveal blind spots, undercounted sources, or abnormal releases that inventories smooth over. Models are useful because they help reconcile scattered evidence and infer broader patterns that no single observational system can provide alone. Each tool has a job. The trouble begins when a tool is asked to do the whole job by itself.

The politics of baselines is therefore not merely a matter of denial versus honesty. It is often a matter of institutional incentives. Governments want internationally defensible numbers. Regulators want thresholds they can enforce. Operators want methods that are stable enough to plan against. Scientists want estimates that better reflect observed reality. Finance ministries want figures solid enough to justify spending. Diplomats want claims that will survive comparison across jurisdictions. These demands overlap and also collide. A baseline generous enough to preserve reporting continuity may miss costly leaks. A baseline revised too abruptly may trigger claims that prior policy was built on sand. The number moves, and with it the politics.

Credibility is where these tensions become strategic. Public trust does not survive long on the vague promise that experts are handling things. It survives when institutions can explain why different methods yield different answers without implying that everything is up for grabs. That is harder than it sounds. A government that minimises disagreement may look evasive when new measurements surface. A government that overstates uncertainty may sound as though it lacks control. Credibility in this field comes from procedural clarity: what data are being used, what their limits are, how discrepancies are investigated, how revisions are made, and what decisions can still proceed before every method converges. Trust is visible here. It appears in whether reporting rules are updated, whether monitoring budgets are maintained, whether ministries revise estimates openly, and whether international partners continue to treat a country's figures as serious.

Budgeting is tied to the same problem. A baseline is not only a scientific statement; it is an anchor for spending. Inspection regimes, repair programmes, monitoring networks, diplomatic commitments, infrastructure adaptation, and emergency planning all depend on some judgment about the scale and distribution of methane-related risk. If estimates shift, budgets may need to shift as well. That can be politically painful. Governments prefer expenditure cases that rest on settled quantities. Methane systems often deny that comfort. A treasury asked to fund more monitoring, tighter regulation, or revised infrastructure planning may push back not only because it doubts the hazard but because it doubts the stability of the underlying numbers. Baseline uncertainty thus becomes a fiscal fact.

International coordination raises the stakes further. States do not compare methane performance through a single universal instrument. They compare reported inventories, sectoral rules, independent monitoring, research findings, and increasingly public satellite-based observations. If one country reports a stable emissions profile while remote sensing repeatedly identifies strong point sources, diplomatic trust frays. If another country revises its baseline upward after improved measurement, it may be penalised politically for being more transparent than a less candid peer. This creates a perverse incentive structure. Good measurement can look like worsening performance. Weak measurement can preserve appearances. International coordination becomes harder when the baseline is not just contested but strategically awkward.

And yet a limitation must be acknowledged. Not all methane disputes are profound contests over reality. Some differences between methods are small in practical terms for a given decision. Some sectoral baselines are good enough to support immediate action even if they are not perfect in every detail. Some emissions reductions can be pursued sensibly without waiting for full methodological reconciliation. It would be a mistake to turn epistemic complexity into an argument for paralysis. Policy does not require metaphysical certainty. It requires disciplined enough knowledge to act proportionately and revise when warranted. The reconciliation lies there: contested baselines should slow overconfidence, not action itself.

This becomes especially important when methane is discussed as a strategic variable rather than merely an emissions category. The state's problem is not only to know the exact number. It is to decide whether infrastructure assumptions remain sound, whether monitoring systems justify stronger maintenance cycles, whether remote regions are becoming costlier to serve, whether liability exposure is increasing, and whether public claims about preparedness still rest on credible evidence. A baseline, in that context, is part measurement and part administrative commitment. It tells institutions what they think they are governing.

A quiet but telling motif runs through all of this: the reading. A reading from a field sensor. A reading from a satellite pass. A reading in an emissions inventory table. A reading taken from a model output. Each seems to promise contact with the real. Each does make contact, but never from nowhere. It comes through calibration choices, temporal windows, spatial coverage, assumptions about behaviour, and institutional purposes. The reading changes. A mature security analysis does not panic at that fact, nor does it hide it. It asks what kind of decision can still be defended when the reading changes and what kind of institution is able to say so plainly.

The most serious error is to imagine that contested baselines are a temporary embarrassment on the way to a final stable picture. Methane systems are dynamic, and the methods used to observe them are changing as well. Better tools will sharpen some disputes and settle others. New observations may reveal undercounted sources, improve sectoral understanding, or complicate earlier assumptions about natural variability and sinks. That is not a sign of failure. It is the condition of governing a difficult subject honestly. The hard question is what governments will do with that honesty. Will they treat methodological disagreement as a reason to postpone difficult commitments, or as a reason to build budgets, reporting systems, and international claims that can withstand revision without collapsing into mistrust?

Arctic amplification and permafrost relevance

In May 2020, a diesel tank at a power plant near Norilsk in northern Siberia collapsed and spilt fuel into rivers and tundra. The operator was NTEK. The place was the Russian Arctic. The date is easy to verify. Public attention fixed, understandably, on the spill itself: polluted water, damaged land, a costly emergency response. Yet the deeper strategic lesson lay under the tank. Ground, once treated as a reliable foundation, had become less reliable. The floor had shifted.

That image matters because Arctic change is often discussed at the wrong altitude. From distant capitals, the high north is frequently imagined through broad categories - ice, shipping routes, mineral prospects, military presence, geopolitical competition. All are real enough. But the most consequential changes often begin in something more prosaic: the behaviour of the substrate beneath roads, depots, runways, storage sites, pipelines, housing, and communications masts. Permafrost is not merely frozen ground in a scientific description. In public administration, it functions as load-bearing expectation. When it thaws unevenly, softens seasonally, or becomes harder to predict across space and depth, the state is forced to govern not just territory, but uncertainty about what territory can still support.

Arctic amplification draws strategic attention even from governments far from the polar circle because the Arctic is not an isolated theatre. It is a zone where physical change intersects with shipping, resource extraction, military infrastructure, indigenous settlement, scientific monitoring, and national prestige. More importantly, it is a region whose apparent remoteness hides a dense web of dependencies. Remote radar sites rely on stable access. Offshore platforms rely on emergency reach. Northern communities rely on runways, seasonal roads, fuel depots, and communications links whose failure alters sovereignty in practice, not only in rhetoric. Distant capitals care because the Arctic tests whether a state can still perform ordinary functions where the physical basis of those functions is changing faster than administrative habits tend to admit.

The phrase "Arctic amplification" can sound abstract when left in atmospheric language. Its practical meaning is simpler and harsher. Changes in temperature and related environmental conditions at high latitudes affect ice, snow, hydrology, vegetation, erosion, and the freeze-thaw behaviour of the ground. Permafrost relevance follows from that chain. Where infrastructure, storage, or housing relies on ground remaining sufficiently stable within expected ranges, uncertainty about the ground becomes uncertainty about the asset. The concept to name here is substrate risk. It is not dramatic enough for political speeches. It is exactly dramatic enough for maintenance budgets, insurance decisions, engineering schedules, and emergency planning.

Show the issue at the facility level, and the strategic meaning sharpens. A remote fuel depot is inspected on a normal cycle. A runway is resurfaced according to expected wear. A pipeline support is monitored under assumptions about the seasonal rhythm of the ground. A housing block is repaired after visible settling. These tasks look routine. Once thaw depth, drainage patterns, and soil stability begin to shift, routine becomes deceptive. Inspection intervals may no longer match the pace of deterioration. Surface repairs may mask deeper movement. Seasonal access windows for heavy equipment may shorten. Costs rise. Confidence drops. The implication is not simply that assets become more expensive. It is that planning assumptions lose their quiet authority.

That loss of authority helps explain why Arctic change commands attention from defence ministries, finance departments, transport agencies, and local governments at the same time. Security institutions do not worry about permafrost because they have developed an aesthetic taste for frozen landscapes. They worry because ground instability can degrade runways, roads, fuel storage, radar installations, ports, housing, and emergency access in places where redundancy is sparse and repair is difficult. Fiscal institutions care because deferred maintenance, emergency repair, relocation, and liability exposure all alter budget trajectories. Municipal and regional governments care because ordinary public services become harder to deliver when the built environment no longer behaves as expected. One physical process enters several ledgers at once.

A named example from Alaska shows how these pressures operate without requiring spectacular collapse. The Dalton Highway and infrastructure associated with the North Slope have long forced engineers and operators to contend with difficult ground conditions, frost heave, drainage problems, and the need for continual maintenance. The relevance of this example is not that Alaska offers a perfect model, or that all Arctic systems look alike. It is that serious operators in high-latitude environments have long known that the ground cannot be treated as an inert backdrop. The strategic lesson is plain. Where institutions assume that adaptation can be left to occasional repair, they will be surprised. Where they treat maintenance as a standing function of governing unstable terrain, they perform better, though often at high cost.

That cost matters because the relevance of permafrost is not confined to major industrial corridors or high-profile national assets. It reaches smaller settlements and less visible facilities whose failure may be politically quieter and socially harsher. A deformed runway in a remote community is not only an engineering issue. It can affect medical evacuation, food prices, school attendance, mail, policing, and the practical sense that the state still arrives when needed. A settling foundation under housing stock is not only a property problem. It can alter insurance availability, municipal finance, household debt, and decisions about whether younger residents stay or leave. Strategic significance does not always announce itself with tanks or treaties. Sometimes it appears in missed flights and cracked slabs.

This is where permafrost becomes relevant to methane in a specific rather than merely atmospheric sense. Public debate sometimes jumps too quickly from thawing ground to dramatic claims about massive methane release and runaway effects. Serious analysis should resist that leap. The planning problem does not require sensational prediction. It requires acknowledgement that high-latitude warming and associated ground instability already matter for the integrity of built assets, and that these changes occur in environments where methane-related feedback concerns have sharpened attention to timing and uncertainty. One need not predict exact future releases from thawing permafrost to recognise that permafrost dynamics are strategically relevant now because the substrate supporting existing infrastructure is becoming harder to trust.

The temptation to overstate is real, and it deserves to be checked. Not every failure in the north is caused by thaw. Some infrastructure was poorly designed or badly maintained from the outset. Some communities live with chronic underinvestment unrelated to present climatic shifts. Some assets have long required expensive maintenance because Arctic environments have always been demanding. A road washout or foundation crack is not self-proving evidence of climate-driven transformation. This is a genuine limitation on the argument. It prevents the analysis from turning every northern difficulty into a morality play about warming. The stronger claim is narrower: when ground conditions become less predictable across time and place, the threshold for prudent planning changes even where causation at the individual asset level remains mixed.

That narrower claim is also more useful. It directs attention away from theatrical certainty and towards administrative consequence. A ministry does not need to prove that every millimetre of ground movement at a given site resulted from a specific climate mechanism before adjusting maintenance cycles. A facility manager does not need a perfect model of future thaw depth before asking whether inspection routines, repair contracts, access logistics, and replacement parts are calibrated to the wrong tempo. The issue is not omniscience. It is whether institutions can recognise that substrate reliability has become a variable rather than a constant.

Maintenance cycles offer the clearest way to see this. Under stable conditions, long-lived facilities can be serviced at established intervals informed by past wear patterns, known access windows, and predictable environmental loads. In less stable conditions, those intervals themselves become risky assumptions. A remote communications mast may require more frequent foundation inspection because subsidence can develop unevenly. A fuel storage site may need redesigned monitoring because surface integrity no longer guarantees sub-surface stability. A runway may need earlier intervention because freeze-thaw behaviour and drainage have changed in ways not captured by the original maintenance schedule. Maintenance is no longer merely upkeep. It becomes a form of reconnaissance.

That shift has political consequences. Preventive maintenance is easy to postpone because its success appears as non-event. Yet in Arctic systems, delay can be especially costly because repair seasons are short, transport is difficult, and replacement capacity is often far away. A deferred intervention may not simply become more expensive next year. It may become impossible within the next workable season, forcing emergency contracting at greater cost and under worse conditions. The strategic point is not that every northern asset now requires constant alarm. It is that the old administrative comfort of routine intervals and deferred attention weakens when the ground itself becomes part of the uncertainty.

This is also why distant capitals cannot treat Arctic change as a matter for local authorities alone. Permafrost relevance cuts across scales of governance. National governments often own, fund, regulate, or insure northern infrastructure directly or indirectly. They bear responsibility for defence sites, transport corridors, emergency services, environmental liability, and sovereign presence expressed through service provision. Even where municipal or regional actors handle day-to-day operations, the fiscal and political consequences of failure travel upward quickly. A repeated service outage in a remote region can become a national question about competence, fairness, and territorial commitment. Arctic amplification may be a climatic phrase. Its administrative consequences are constitutional in feel.

A second named example points in the same direction. In northern Canada, the reliability of winter roads has long shaped the economics and public services of remote communities and industrial sites. These roads rely on cold, stable conditions that allow heavy transport across frozen terrain and lakes during limited seasons. When those conditions become less reliable, the issue is not simply a matter of transport inconvenience. Fuel deliveries, construction schedules, food supply, mining logistics, and public budgets are all affected. The relevance to permafrost is not identical in mechanism, but it is analogous in governance. A background condition once treated as seasonally dependable becomes a planning variable. States do not experience that as meteorology alone. They experience it as pressure on procurement, stockpiling, cost control, and service credibility.

Monitoring complicates matters further. Ground temperature measurements, remote sensing, site inspections, and engineering assessments can improve awareness of changing conditions. They cannot abolish uncertainty. Permafrost is heterogeneous. Local drainage, vegetation, topography, soil composition, snow cover, and built load all matter. A regional map or broad projection may inform policy while saying little decisive about the next failure point at a particular facility. This is why Arctic planning so often frustrates central institutions. They want thresholds. The substrate offers gradients, patchiness, and changing confidence. Better information helps, but only when institutions are willing to revise standards, budgets, and maintenance practices before every ambiguity is resolved.

Trust enters here in a stubbornly practical way. Local residents, engineers, facility managers, and indigenous observers often notice changes in ground behaviour, drainage, or seasonal timing before these appear in central reports. If institutions ignore such observations because they do not arrive in the preferred technical format, they lose not only goodwill but also a signal. If they rely on anecdote without disciplined comparison, they risk acting on noise. Capability lies in combining situated observation with formal monitoring, then tying both to authority and resources. Trust, in this setting, is visible when early warnings travel upward without ridicule and when revised maintenance or relocation decisions travel downward without euphemism.

A fair counter-example should be kept in view. Some Arctic infrastructure performs better than public debate assumes because it was designed with difficult ground conditions in mind, monitored intensively, and supported by substantial maintenance budgets. Elevated pipeline segments, engineered foundations, and adaptive maintenance regimes can and do reduce risk. This matters because it prevents a lazy slide from "permafrost relevance" to "inevitable collapse". High-latitude change is not a uniform sentence passed upon all northern assets. But the counter-example also clarifies the burden. Good performance usually depends on sustained technical attention, funding, and institutional seriousness. It is not evidence that the underlying problem is overstated. It is evidence that the problem can be managed only by treating substrate uncertainty as a permanent planning condition.

That condition forces a different way of reading strategic significance. The Arctic is often framed as a frontier of opportunity or rivalry. Both framings capture something real. They also risk missing the quieter determinant of success or failure: whether the ground under critical systems remains dependable enough for budgets, maintenance routines, and service promises to hold. Sovereignty is not only asserted by maps and patrols. It is experienced through runways that stay open, fuel that remains safely stored, roads that support resupply, and buildings that continue to stand where the state said they would.

The floor had shifted in Norilsk, but the larger significance of that image is broader than one spill. It suggests a test that reaches from industrial sites to village runways, from national defence planning to municipal repair budgets. How much uncertainty about the substrate can a state absorb before it is no longer merely maintaining Arctic presence, but improvising it from one expensive exception to the next?

Satellites, sensors, and governance

In January 2022, the European Space Agency drew wide attention when data associated with one of its satellites helped identify a major methane release near an underground gas storage facility close to Los Angeles. The institution was named. The date is easy to verify. The place was visible enough on a map. The public lesson seemed straightforward: the sky can now see what ground systems miss. Yet the deeper lesson was less triumphant. Detection is not action. A plume on a screen does not repair a valve, authorise an inspection, settle liability, or decide how quickly the public should be told what is known and what is still uncertain. The signal arrived. The pathway mattered.

That distinction is the heart of monitoring under climatic stress. Satellites, aircraft, ground sensors, thermal systems, meteorological stations, permafrost probes, sea-ice observation platforms, and infrastructure monitoring devices can all widen institutional vision. They can show trends. They can provide alerts. They can narrow the range of plausible causes after an event. They can reveal blind spots in national inventories or operator reporting. What they cannot do by themselves is govern. Governance begins when information is tied to authority: who must be told, who may decide, what threshold triggers what response, how uncertainty is recorded, and how the inevitable gap between signal and certainty is handled without drift or panic.

Public argument often swings between two fantasies. One imagines that more data will dissolve political conflict, as though every monitoring gain naturally becomes shared understanding and timely action. The other imagines that because data are imperfect, they are politically disposable until certainty arrives. Both are wrong. Monitoring systems are indispensable, but they are never neutral instruments floating above institutions. They are embedded in budgets, mandates, software standards, reporting cultures, legal authorities, procurement cycles, and public expectations. A sensor is not merely a device. It is part of an administrative chain. If that chain is weak, more sensing can produce more frustration rather than more control.

Start with what these systems can realistically provide. The first and most basic contribution is trend detection. Repeated observation over time can show whether methane concentrations, leak patterns, thaw depth, ground movement, or ice conditions are changing in ways that matter for planning. Trend information is rarely dramatic. It is exactly the sort of evidence that should inform maintenance cycles, budget assumptions, inspection priorities, and longer-term capital decisions. A single reading may be noisy. A pattern is harder to dismiss. Yet even trends require judgement. What counts as a sufficient signal to revise a standard, expand monitoring, or fund redesign? Without pre-agreed answers, trend data accumulate while institutions hesitate.

The second contribution is alerting. Some monitoring systems are valuable because they can indicate anomalous conditions quickly enough to trigger inspection, warning, or emergency response. Satellite methane observations can sometimes identify unusually large point sources. Ground sensors can detect subsidence, instability, or dangerous concentration levels near infrastructure. Remote sensing can indicate unusual heat, thaw, or changing hydrology around critical assets. Alerts are politically seductive because they feel immediate. They promise operational relevance. But an alert without a response protocol is often little more than a public embarrassment waiting to happen. If no agency knows who owns the next step, or if the threshold for action is contested after the alert arrives, speed becomes theatre.

The third contribution is attribution range. Monitoring rarely delivers courtroom certainty at once. What it can often do is narrow the plausible set of causes or responsible zones. A satellite image may indicate that a release came from one facility cluster rather than another. A network of sensors may show that ground instability is increasing across a site, even if the exact failure point is not yet known. A combined reading from meteorological data, field inspection, and remote sensing may distinguish between an isolated equipment problem and a broader pattern of operational weakness. Attribution range matters because decision-makers rarely need absolute certainty to do their job. They need enough certainty to justify inspection, temporary restrictions, warning messages, contingency deployment, or revision of assumptions. The signal narrows the field.

This is where governance enters decisively. Information becomes actionable only when institutions decide in advance what different classes of information mean. A faint but repeated signal may justify more monitoring. A strong and isolated alert may justify immediate inspection. A repeated anomaly near critical infrastructure may justify revised maintenance intervals, temporary shutdown, or public notification. None of these choices can be improvised well under pressure if authority is unclear. A ministry may have the data but not the enforcement power. A regulator may have the mandate but not the technical staff. A local authority may observe a problem and still need national permission to act. Monitoring systems therefore function as institutional infrastructure: useful only when embedded in pathways that convert observation into decision.

A clear example came in September 2022, when the Nord Stream pipeline leaks in the Baltic Sea were detected through a combination of seismic signals, pressure changes, and surface observations. The geopolitical circumstances were extraordinary, and the event does not map neatly onto methane governance in peacetime civil settings. Still, it offers a useful lesson in information sequence. Detection did not emerge from one magical sensor. It emerged through multiple systems that together established that something serious had occurred. Just as important, the event showed how quickly data become politicised when attribution, liability, and strategic meaning remain contested. Monitoring clarified part of the reality and intensified the argument over the rest. That is not a failure of monitoring. It is a reminder that information enters politics through institutions already under stress.

A second example, closer to ordinary planning, comes from permafrost and infrastructure monitoring in northern regions. Ground temperature measurements, borehole data, satellite-based deformation tracking, site inspections, and local observations can together indicate whether the substrate beneath roads, runways, storage sites, or housing is becoming less stable. None of these streams is sufficient alone. Satellite-based deformation data may indicate movement without explaining exact subsurface cause. On-site inspection may reveal visible damage after a broader trend has already been underway. Local users may notice drainage changes or surface irregularities before formal systems register them. The governance lesson lies in integration. Monitoring succeeds not when one instrument wins the argument, but when institutions know how to compare signals, escalate concern, and adjust practice before failure becomes the first unambiguous proof.

That requirement sounds technical. It is also constitutional in a small-c sense because it concerns who may decide under uncertainty. Decision pathways are often the missing link in public preparedness. Agencies collect data because collection is fundable, visible, and defensible. Far fewer institutions invest comparable effort in specifying which official can authorise road closure, facility inspection, emergency procurement, public warning, temporary curtailment, or budget reallocation on the basis of imperfect but credible information. The result is a quiet pathology of contemporary governance: better sensing paired with unchanged delay. Officials know more, earlier, and still move too late because the authority structure was built for calmer conditions and narrower questions.

This is one reason data can increase politicisation when decision pathways are absent. More information raises expectations. Journalists, local communities, opposition parties, investors, and international partners all assume that a visible signal should produce visible action. If action does not follow, suspicion grows. Was the data flawed? Was the institution hiding something? Is the agency captured by the regulated sector? Is uncertainty being used as cover for inaction? These suspicions may or may not be fair in a given case. What matters analytically is that monitoring changes the political environment. It does not merely add knowledge. It alters the burden of explanation. A state that builds sensing capacity without building decision capacity can look more informed and less competent at the same time.

Trust becomes concrete here. It is not a noble sentiment hovering above the technical apparatus. It appears when local officials report anomalies because they expect a serious response, when regulators publish updates because they believe transparency will not be punished automatically, when communities heed warnings because previous warnings proved operationally useful, and when operators accept inspection triggers because the rules were defined in advance rather than improvised opportunistically. Trust in monitoring systems is built through visible behaviour: calibrated instruments, clear reporting intervals, public correction of errors, consistent thresholds, cross-agency information sharing, and the boring discipline of follow-through. Without these habits, every alert risks being treated as either overreaction or concealment.

A common error is to imagine that more granular data always produce better governance. Sometimes they do the opposite. A flood of readings can overwhelm staff, generate contradictory interpretations, or encourage decision-makers to wait for one more update before acting. High-resolution information can also tempt institutions into false confidence, as though detailed observation has abolished the uncertainty that still surrounds consequence and response. Precision in measurement is not the same as clarity in decision. An agency may know exactly where methane concentration spiked and still have no settled rule for what happens next. A local authority may receive frequent ground movement data and still lack the budget or legal mandate to accelerate repairs. More pixels do not solve governance.

That limitation should be acknowledged plainly. Some monitoring systems remain partial, intermittent, or expensive. Satellite detection depends on orbital coverage, atmospheric conditions, instrument sensitivity, and the nature of the source. Ground sensors can fail, drift, or cover too little terrain. Models can reconcile data streams and still depend on assumptions that remain contestable. A sceptic could look at this and conclude that monitoring is too uncertain to deserve strategic weight. That would be a mistake. The correct lesson is narrower. Monitoring cannot deliver omniscience. It can materially improve institutional timing, comparison, and prioritisation when designed to feed decisions rather than decorate reports. Imperfect information can still be highly actionable if the organisation has agreed how to use it.

This is why mandates matter so much. If one agency monitors, another inspects, a third funds repair, and a fourth communicates with the public, then the monitoring system is only as useful as the coordination among them. Fragmentation is common. Environmental agencies may hold technical expertise but lack operational authority. Infrastructure ministries may control assets but rely on external scientific advice they do not fully understand. Regional governments may face the consequences of delayed action while national authorities retain key powers. In such systems, data often move faster than decisions. Monitoring then becomes politically dangerous because it reveals problems that the administrative structure cannot yet absorb.

Thresholds matter just as much. They are often misunderstood as purely scientific cut-offs, discovered rather than chosen. In practice many thresholds are governance commitments. Institutions decide that a given level of anomaly, uncertainty, or trend change will trigger inspection, maintenance, warning, or funding because waiting longer would expose them to unacceptable risk. These thresholds should be informed by science, but they cannot be derived from science alone. They reflect tolerance for disruption, budgetary constraints, legal duties, and political appetite for preventive action. A mature monitoring regime states this honestly. It does not pretend that every trigger descends directly from nature.

Response protocols complete the chain. A protocol does not eliminate uncertainty; it distributes it. It says, in effect: if signal X appears under conditions Y, institution Z will do A, B, and C while uncertainty remains at level D. This matters because climate-linked risk often unfolds faster than interagency bargaining can manage. Protocols reduce the number of decisions that must be invented under stress. They also make later review possible. One can ask whether the threshold was set too high, whether escalation was too slow, whether communication was too vague, whether maintenance followed the signal properly. Without protocols, every crisis becomes both event and organisational audition.

A fair counter-example lies in settings where strong monitoring systems appear to work well precisely because institutions have already normalised these linkages. Some flood management regimes, certain air traffic systems, and parts of industrial safety monitoring show what disciplined information-to-action chains can look like. The lesson is not that methane or permafrost monitoring can be copied wholesale from those domains. The hazards differ. The institutional environments differ. The useful point is that monitoring works best where thresholds, authority, and revision habits are treated as part of the infrastructure rather than as afterthoughts. A sensor network alone is not the model. A governing arrangement is.

For methane systems and high-latitude change alike, the real strategic shift comes when monitoring is no longer seen as the collection of data points but as the maintenance of institutional sight. Sight, however, is not the same as judgment. A state can see a great deal and still fail because it cannot decide quickly, speak clearly, or act across organisational seams. That is why the public often experiences monitoring indirectly: not as maps or readings, but as whether a road closes before collapse, whether a facility is inspected before a major release, whether evacuation guidance changes in time, whether repeated anomalies alter maintenance budgets, whether officials admit what they do not yet know without sounding absent or evasive.

The signal arrived near Los Angeles in 2022, and the larger lesson survives that one event. Modern institutions can observe more than their predecessors could. They can track atmospheric anomalies, ground movement, thermal irregularities, changing ice, and infrastructure stress with remarkable sophistication. The unanswered question is harder and more political. When the next alert appears - not as a perfect answer, but as a credible warning inside a cloud of uncertainty - will the state treat the signal as information to govern by, or as another inconvenient reading waiting for someone else to own the pathway?

Communicating uncertainty without paralysis

In August 2023, officials in the Northwest Territories ordered the evacuation of Yellowknife as wildfire threatened the city. The place was northern Canada. The date is easy to verify. The hazard itself was frightening enough, but the communication problem was no less serious: forecasts evolved, smoke complicated movement, roads and flights became matters of urgency, and public guidance had to change without sounding either evasive or erratic. People were not listening for perfect certainty. They were listening for whether the next instruction deserved to be trusted. The message moved.

That scene captures a central difficulty in climate-linked security planning. Uncertainty is often heard in public life as a confession of ignorance, incompetence, or manipulation. If officials say they are not yet sure, some citizens hear that nothing is known. If they revise a warning, others hear proof that the original message was misleading. If they speak too firmly and later adjust, they can appear deceptive. If they hedge too much, they can appear absent. This is not merely a communications challenge in the narrow sense. It is a problem of statecraft under pressure. The issue is whether institutions can speak with enough candour to remain credible and enough clarity to keep people moving, complying, and cooperating while the picture is still changing.

The first task is conceptual. Not all uncertainty is the same, and public language becomes much more useful when it reflects that difference. A *known unknown* is a matter that institutions recognise as unsettled but bounded. Officials may know that a fire could shift with wind, that thaw-related instability may affect a facility without yet knowing the exact point of failure, or that a methane reading indicates a likely problem without yet fixing the full scale or source. The range of possibilities is real, though incomplete. An *unknown unknown* is different. It refers to conditions, interactions, or failure modes that were not built into the plan at all. A backup system depends on a supplier whose own vulnerability was never considered. An evacuation route remains open, but a receiving facility fails. A regional mutual aid agreement collapses because every neighbouring jurisdiction is also under strain. Known unknowns can be discussed in advance. Unknown unknowns are discovered by contact.

This distinction matters because it changes how institutions should speak. Known unknowns invite conditional communication: what is known, what is not yet known, what is being watched, and what action follows under different developments. Unknown unknowns demand a different tone - one that acknowledges the possibility of surprise without making surprise sound like abandonment. Public reasoning becomes stronger when people are not asked to choose between false certainty and fatalism. A warning can be both provisional and serious. A plan can be both revised and credible. That is the governing skill.

Show the opposite and the stakes become plain. After severe weather, fire, or flooding, officials sometimes delay strong public messaging because they fear being accused of alarmism if the worst does not happen. That hesitation may seem prudent inside an office. Outside it, delay can shorten evacuation windows, crowd roads, increase competition for fuel, and leave care facilities scrambling to interpret what the state actually expects them to do. The concept to name here is decision-lag in communication. The implication is practical: uncertainty does not only live in models or forecasts. It lives in the shrinking interval during which words still have operational value.

The temptation to smooth uncertainty away is strongest when institutions worry about appearing weak. Here the habits of ordinary political communication can become dangerous. Public life rewards confidence. It punishes hesitation. Ministers, mayors, and agency heads know that an unqualified statement is easier to quote than a careful one. Yet under climate-linked disruption, overconfidence can be more corrosive than visible caution. If a government repeatedly speaks as though conditions are stable and then revises abruptly, citizens do not merely update their weather expectations. They update their judgement of institutional honesty. Trust is then lost not because uncertainty existed, but because institutions denied its place in the message.

COVID-19 provided an unmistakable demonstration of this dynamic, even though the hazard was epidemiological rather than climatic. In many countries, public trust rose or fell not only with the substance of policy but with the relationship between certainty claimed and certainty available. Guidance that changed was not always the problem. Guidance that changed after being presented as settled often was. The lesson transfers directly. Evacuation messaging, smoke advisories, water restrictions, fuel prioritisation, infrastructure warnings, and service continuity updates all require a style of communication that can revise without appearing to confess prior bad faith. That style does not emerge spontaneously in crisis. It has to be built into the institution's habits.

A second example makes the point more concretely in a climate-linked emergency. During Hurricane Sandy in October 2012, public communication in the United States had to move from forecast uncertainty to immediate operational instruction across multiple jurisdictions. The most useful messages did not pretend the storm path, local impacts, or cascading failures were known in perfect detail. They translated evolving information into clear behavioural guidance: where evacuation applied, what transport disruptions were likely, what services would close, what people should not assume would remain available. The issue was not atmospheric precision alone. It was whether communication converted uncertainty into action rather than passivity. Where messaging became confused, delayed, or fragmented, the public experienced not just informational noise but institutional friction.

This is why credible communication should be treated as a security asset rather than a public-relations afterthought. Security assets preserve function under stress. Communication does exactly that when it keeps people from overwhelming fragile systems, helps them sequence decisions properly, and reduces the space in which rumour can substitute for instruction. It shapes whether roads clog too late, whether fuel stations are rushed prematurely, whether hospitals face avoidable surges, whether local authorities are contradicted by national ones, and whether evacuation orders are heard as commands grounded in planning or as bureaucratic panic. Credibility, here, is measurable in behaviour. Do people leave when told? Do institutions share the same wording? Do frontline workers know which update supersedes the last? Does changing guidance increase compliance or collapse it?

The phrase "without paralysis" is therefore important. Some officials avoid discussing uncertainty because they fear that ambiguity will freeze the public. Sometimes it does. More often paralysis results from a different failure: communication that treats uncertainty as too embarrassing to explain and therefore provides little guidance until the situation is already narrow and frightening. Most people can act under conditional information if the conditions are intelligible. "Leave now because fire behaviour may close this route later" is usable. "We continue to monitor the situation closely" often is not. One names a plausible development and a present decision. The other names institutional awareness without telling the public what that awareness should change in their behaviour.

This is where the distinction between known unknowns and unknown unknowns becomes operational rather than philosophical. Known unknowns should be built into messaging structure. Authorities can say, in effect: these are the outcomes we are watching, these are the signs that will change our instruction, and this is what you should do before certainty narrows further. Such messages preserve room for revision because revision has already been introduced as part of the process. Unknown unknowns require a more modest but equally serious signal: that plans exist, but systems can still surprise; that people should expect updates; that the absence of full explanation in the first hour is not proof of concealment; that adaptation is part of competent management. This does not abolish fear. It prevents revision from sounding like improvisation.

Still, communication alone cannot rescue weak institutions. This limitation matters. A beautifully phrased warning cannot compensate for roads that are already impassable, shelters that are full, fuel reserves that were never stocked, or command structures too fragmented to produce coherent decisions. To praise credible communication as a security asset is not to elevate rhetoric above logistics. It is to recognise that logistics themselves depend on the behavioural cooperation that communication helps produce. Good messages can preserve an evacuation window. They cannot invent one. Good public updates can reduce panic buying. They cannot fill empty depots. Honest acknowledgement of uncertainty therefore works only when linked to real decision pathways and visible preparation.

This is why communication must be treated as part of institutional infrastructure. It depends on protocols, delegated authority, training, message discipline, and relationships between agencies. If one office issues a warning while another issues a reassurance, the public receives contradiction as a fact of governance. If national agencies speak in probabilities while local authorities speak in absolutes, confusion widens. If frontline staff cannot tell which version is current, institutional coherence dissolves where citizens meet it directly. Communication systems need maintenance in the same way physical systems do. They require rehearsal under changing conditions, not merely drafting under calm ones. A message is a hand-off.

Rumour and misinformation sharpen the stakes. During disruption, people do not wait quietly for official clarity. They compare texts, social media posts, local observations, and hearsay from family or neighbours. Some of this informal information is useful. Some of it is dangerously wrong. The best defence is not merely fact-checking after falsehood spreads. It is a rhythm of official communication that is frequent enough, specific enough, and honest enough about uncertainty that people do not feel abandoned to private speculation. When institutions disappear into silence because the picture is not yet complete, they create a vacancy that rumour quickly fills. Communication then ceases to be a neutral channel. It becomes part of the hazard environment.

This is particularly important for methane-related systems and monitoring limits because much of what institutions know arrives as trend, anomaly, or attribution range rather than final certainty. A satellite alert may indicate a likely release without settling source and scale. A ground monitoring system may show unusual movement without predicting exact failure. A seasonal pattern may point to deteriorating reliability without dictating the exact day of closure or breakdown. Public communication in such settings must avoid two opposite errors. One is to imply that because the signal is not perfect, there is nothing operational to say. The other is to present preliminary information with more confidence than it deserves. The first breeds delay. The second breeds backlash when revision becomes necessary.

A fair counter-example comes from domains where clear communication under uncertainty has worked comparatively well because institutions normalised conditional updates before crisis struck. Some meteorological agencies and emergency management systems have learned to present forecast cones, warning tiers, and update sequences in ways that teach the public how uncertainty behaves. This is not a complete solution, and even strong systems can fail under compound stress. But the lesson is real. When revision is routine rather than exceptional, trust becomes less fragile. People come to understand that a changed message can indicate increased precision rather than prior deception. The state's credibility then rests less on pretending certainty and more on showing disciplined movement as the picture sharpens.

That movement has an ethical aspect, but it is not primarily moral theatre. It concerns the distribution of burden. Vulnerable populations, care-dependent residents, remote communities, and people with fewer transport or financial options often need earlier and clearer signals than others because their decisions take longer to execute. If institutions communicate as though everyone can wait for the final, most certain update, they quietly favour the mobile and well-resourced. Uncertainty communication is therefore also about fairness in time. Who is given enough warning to act before options narrow? Who is spoken to in ways that match their actual constraints? Whose risk is treated as acceptable collateral of official caution?

A mature public language of uncertainty should therefore do four things at once. It should distinguish what is known from what remains open. It should state what action follows from that imperfect knowledge. It should explain why guidance may change without implying that change itself signals failure. And it should preserve enough institutional humility that new information can be absorbed without reputational panic. None of these habits is easy in political systems built around certainty performances. All of them are compatible with authority. Indeed, authority under climatic stress may increasingly depend upon them.

The message moved in Yellowknife because the hazard moved, the forecasts evolved, and the costs of delay were real. That is no anomaly. It is a preview of the communicative burden climate-linked disruption places on public institutions. The question is not whether uncertainty can be removed from the message. It cannot. The question is whether governments can speak uncertainty in a way that still tells people what to do, when to do it, and why the next update should be read not as a contradiction but as proof that someone is still governing while the picture changes.

Chapter 4

Permafrost Thaw and Infrastructure Exposure

What happens to sovereignty when the road still exists on the map but no longer carries what the state needs it to carry? That question is more exact than it first appears. Infrastructure in high-latitude regions is often discussed as though its significance lay mainly in engineering difficulty or development cost. Yet roads, runways, storage sites, pipelines, and municipal systems do more than move goods or people. They make public authority visible. They are how mail arrives, how patients are evacuated, how schools stay open, how fuel reaches generators, how police and repair crews appear when called. When the ground beneath these systems begins to lose its old reliability, the failure is never only physical. The timetable slips.

Permafrost thaw sharpens this point because it unsettles infrastructure from below. A cracked slab, a tilted foundation, a rut deepening in a road surface, a drainage channel behaving differently than it did a few years earlier - these can look like local inconveniences or routine wear. They are often treated that way in budgets and maintenance logs. The larger problem is cumulative. Small deformations raise transport costs, shorten repair windows, complicate inspections, and increase the odds that one failure arrives while another is still unresolved. A government may respond to each item separately and still miss the pattern joining them. What looks minor at asset level can become strategic at settlement level.

This is why the chapter's real subject is not thaw alone but exposure. Exposure means more than physical proximity to hazard. It describes the degree to which built systems, fiscal commitments, administrative routines, and public expectations are tied to environmental conditions that are becoming harder to trust. A runway built on shifting ground is exposed in one way; a remote clinic dependent on that runway is exposed in another; a municipal budget forced to choose between repeated repair and service continuity is exposed in yet another. The same thaw reaches each through a different path. Infrastructure is where those paths cross.

The central tension therefore lies between apparent durability and actual governability. Infrastructure projects are often financed and described as long-term commitments, symbols of presence, confidence, and permanence. Permafrost thaw introduces a different rhythm. It compresses maintenance cycles, complicates inspection, shortens seasonal access for heavy repair, and increases the penalty for delay. Public institutions, however, usually move at the pace of contracting rules, staffing shortages, annual appropriations, and procurement lead times. The result is not simply higher cost. It is a tempo mismatch between environmental change and administrative throughput. The spreadsheet lags behind the ground.

That mismatch matters because local legitimacy is built from repeated contact with ordinary competence. Residents do not infer state capacity from abstract doctrines. They infer it from whether the fuel arrives before winter tightens, whether the airstrip remains usable when an emergency flight is needed, whether repeated warnings about unstable ground produce visible repair rather than another study, whether a damaged road is treated as a public obligation rather than a remote inconvenience. When these systems falter, prices rise, services narrow, and households begin to reorganise their lives around the expectation that formal provision will be late or partial. Political legitimacy then weakens not through grand ideological rupture but through accumulated practical disappointment.

The stakes extend far beyond local inconvenience. Linear infrastructure such as pipelines, roads, and power corridors carries risk across distance, linking variable ground conditions to inspection burdens, response delays, and wider confidence in safety regimes. Damage in one place can trigger caution, cost escalation, or diplomatic anxiety elsewhere because investors, insurers, regulators, and neighbouring states read infrastructure reliability as a proxy for institutional seriousness. The same is true in finance. Risk moves from cracked surfaces and unstable foundations into premiums, exclusions, municipal borrowing pressure, and the quiet expansion of public backstops. Fiscal systems often register strategic strain before governments are willing to name it openly.

The path ahead follows that chain where it actually runs. It begins with the mundane systems that make remote settlement viable and state presence credible, then moves into the longer and more brittle infrastructures whose failures can cascade across sectors and jurisdictions. From there the argument settles on maintenance, procurement, and the shrinking room between known need and administrative delivery, before turning to the authority and insight held locally, especially where indigenous governance and lived observation reveal blind spots that central systems miss. It closes that movement in the financial language through which infrastructure risk often returns to politics even when politics has tried to keep the issue technical.

What gives these themes their coherence is a simple but demanding proposition: infrastructure exposure is never just about assets. It is about the fit between assets, institutions, and place. A road can be repaired yet still remain strategically weak if contracts arrive too late, spare parts too slowly, and local warnings too low in the hierarchy to alter plans. A pipeline can be physically sound today and still sit inside a deteriorating confidence environment if inspection limits, response distance, and liability uncertainty begin to widen. A community can possess nominal service coverage and still experience withdrawal if every disruption becomes harder to reverse than the last. Exposure is therefore as much administrative as material.

This is also why the chapter cannot treat local knowledge as an ornament appended to expert systems. Where permafrost conditions shift unevenly, those who live with the terrain often notice emerging hazard in drainage, access, surface movement, and seasonal timing before the formal apparatus catches up. The strategic issue is not whether such knowledge should be honoured symbolically. It is whether institutions can use it without merely extracting it, whether authority follows information closely enough to matter, and whether consultation remains tied to action rather than fatigue. Legitimacy and effectiveness meet at precisely that point.

The harder question arrives before any individual case does. When the substrate beneath a state's ordinary functions becomes less reliable, what counts as a successful response: repairing each visible failure, preserving the service that failure threatens, or admitting that some systems were built on assumptions the administrative machine can no longer keep pace with? The answer will determine whether thaw is treated as a maintenance problem with political consequences, or as a political problem that first reveals itself through maintenance.

Roads, runways, and settlement stability

In August 2023, the Northwest Territories ordered the evacuation of Yellowknife as wildfire threatened the city. The place was northern Canada.

The date is easy to verify. The immediate danger came from fire and smoke, not permafrost. Yet the episode remains instructive for a chapter on thaw because it showed, with unusual clarity, what settlement stability really depends on. It depends on routes that hold, aircraft that can land, fuel that can move, and a public that believes the system will function under strain. The runway mattered.

That simple fact is often obscured by the way remote infrastructure is discussed. Roads and runways tend to appear in policy papers as assets, line items, or engineering problems. For the people who depend on them, they are something more intimate and more political. A road is school attendance, grocery cost, building materials, ambulance access, policing presence, mail delivery, and the ordinary possibility of leaving or returning without drama. A runway is not merely tarmac on compacted ground. It is medical evacuation, specialist care, court access, public service rotation, air freight, emergency response, and the state's promise that distance will not become abandonment. When thaw-related instability reaches these systems, sovereignty stops being an abstract doctrine and becomes a question of whether the next plane can touch down safely.

This is why small failures matter so much. A dip in a road surface, a soft shoulder, a drainage problem, a deformed apron, a roughened runway section, a culvert that no longer drains as expected - none of these looks strategic in isolation. Each can be entered into a maintenance log, assigned a repair category, and treated as routine wear. The concept to name here is cumulative fragility. Thaw does not need to produce spectacular collapse to alter settlement stability. It works more quietly. Repeated minor failures increase travel time, restrict vehicle weight, delay resupply, raise maintenance costs, and shorten the margin for emergency action. The implication is severe: what looks local and technical at the point of damage can become systemic and political at the point of use.

The physical mechanism is straightforward enough. Permafrost and seasonally frozen ground have long provided part of the stable substrate on which northern roads, runways, buildings, and utility systems were planned. As ground conditions become less predictable, surfaces can deform unevenly, drainage can change, and freeze-thaw cycles can produce patterns of stress that older designs and maintenance schedules do not handle well. But the strategic meaning lies not in the mechanism alone. It lies in dependency. A community with one runway or one access road is not simply exposed to hazard. It is organised around that route. If the route becomes less dependable, everything attached to it begins to absorb uncertainty: household planning, public finance, school logistics, police response, food pricing, and the ability of the state to keep its own promises.

Northern Canada offers a clear illustration through the role of winter roads in remote communities. These routes have long functioned as seasonal supply corridors for fuel, food, building materials, and equipment. Their value is not romantic. It is fiscal and administrative. When winter roads open later, close earlier, or operate under tighter load restrictions because conditions are poor, communities do not experience this as a minor transport inconvenience. They experience it through higher prices, altered construction schedules, deferred maintenance, tighter fuel planning, and reduced room for public services to improvise. The road itself may still exist on official maps and annual planning documents. What changes is its reliability. Reliability is the hidden constitutional term.

This matters because settlement stability is not measured only by whether a place remains inhabited. A settlement can remain populated while becoming progressively harder and more expensive to govern. Mail arrives less predictably. Skilled staff rotate less easily. Medical referrals become more fragile. Policing and emergency response lose time. Retail prices rise because freight options narrow. Contractors charge more because access is uncertain. Families adapt privately through stockpiling, earlier purchasing, delayed travel, or eventual relocation. Public systems adapt more slowly because budgets, procurement, and staffing rules do not change at the pace of thaw. The state then appears not as a guarantor of continuity but as a late and costly participant in a problem residents have already been managing themselves.

Runways sharpen this point because they compress dependency into a single strip of maintained ground. In many northern communities, air access is not supplementary. It is decisive. It carries medical evacuation, specialist appointments, urgent freight, public servants, repair crews, and often the practical possibility of emergency exit. A runway does not need to close completely to become strategically weak. Surface deformation, softening margins, drainage issues, or repeated restrictions can reduce the size of aircraft able to operate, narrow weather tolerance, delay critical flights, or raise operating costs enough to alter service frequency. The issue is not only whether an aircraft can land in fair conditions. It is whether the community can still rely on that access under pressure. The runway mattered again.

A brief example from Alaska shows how persistent this challenge can be. Rural Alaskan communities have long depended on airstrips and short runways for transport, medevac, mail, and resupply in environments where roads are absent or seasonal conditions are severe. Reports over many years have linked runway and airfield maintenance challenges to erosion, thaw, drainage problems, and unstable ground in parts of the state. The relevance of this example is not that every airstrip problem is driven by climate change or permafrost alone. It is that air access in remote settings is inseparable from substrate reliability, and substrate reliability is becoming harder to take for granted. A runway in distress is not merely an engineering maintenance item. It is a question about the continuity of public life.

Medical evacuation makes the stakes unavoidable. When a road washes out or a runway is restricted, the immediate concern is often framed in transport terms. Yet transport is only the first link in the chain. A delayed medevac changes clinical risk. A missed specialist transfer alters care outcomes. A late delivery of oxygen, fuel, or replacement equipment can degrade a local clinic's functioning before any dramatic emergency occurs. If a deformed runway limits aircraft weight or operating conditions, emergency planning shifts from "can we fly?" to "under what constraints, with what delay, and who absorbs the risk while the system recalculates?" That is a different order of vulnerability. It is administrative, not only physical.

Policing and routine state presence reveal the same pattern from another angle. Sovereignty in remote regions is not asserted by speeches alone. It is experienced when officers, teachers, nurses, engineers, and maintenance teams can arrive, stay, and rotate reliably. Roads and runways make that possible. Once access degrades, the public may still hear official assurances of equal citizenship while observing a different reality: slower response times, postponed repairs, staff shortages, more expensive goods, and increasing dependence on local improvisation. Political legitimacy erodes through these gaps. The state sounds present and behaves intermittently. That discrepancy can persist for years before it is named as a strategic problem, which is precisely why it is dangerous.

The temptation in such discussions is to look for a dramatic break point - the bridge collapse, the runway closure, the settlement abandonment order. Those moments matter, but they can distort the underlying process. Most communities experience infrastructural decline as a series of minor degradations and awkward workarounds. A gravel road requires more grading. A culvert replacement is delayed. Certain loads are postponed.

A flight is rescheduled. A maintenance contract slips because equipment cannot be moved at the right time. Another repair season is lost to procurement delay. The language of emergency is often absent. The reality is attritional. Costs rise; confidence falls; the public learns to expect less. Settlement stability weakens by instalments.

This is where political legitimacy becomes inseparable from the material condition of infrastructure. Legitimacy is often discussed in broad terms, as trust in institutions or confidence in governance. In northern and remote settings, it is easier to observe. It appears when residents believe an evacuation route will be usable if ordered. It appears that local officials expect that runway defects reported upward will bring timely inspection rather than procedural drift. It appears when prices remain high but not wildly volatile because supply systems still function within known limits. It appears when an emergency call produces a response that feels delayed by weather rather than by indifference. The state earns legitimacy through repeated demonstrations that infrastructure remains governable.

A counter-example is important here. Not all northern roads and runways degrade into failure, and not every settlement facing thaw-related instability slides into service withdrawal. Some jurisdictions have maintained comparatively strong performance through intensive maintenance, redesign, drainage management, and sustained public expenditure. Certain Alaskan and Scandinavian facilities, for example, have benefited from engineering practices and institutional attention that treat unstable ground as a standing condition rather than an occasional exception. This matters because it prevents the analysis from drifting into determinism. Thaw does not make settlement instability inevitable. It makes complacency costly. Where monitoring, funding, and local coordination are serious, some of the worst consequences can be delayed or reduced.

That qualification, however, sharpens rather than softens the political argument. Successful adaptation in such places is rarely cheap or administratively effortless. It often depends on unusual strategic importance, steady maintenance budgets, technical staffing, and infrastructure cultures that accept high upkeep as the price of continued viability. Many remote communities do not enjoy those advantages. They live instead with thin contractor markets, seasonal access limits, staff shortages, and procurement systems designed for standard conditions rather than fast-changing ground. When resilience claims fail there, they usually fail not at the level of engineering theory but at the level of staffing, timing, and parts. The road might be repairable. The repair season is missed. The runway defect might be manageable. The specialist team arrives too late. The formal system knows the answer and still cannot perform it.

This is also why higher prices are not a secondary issue. Price increases are among the first visible signs that physical fragility is entering ordinary life. Freight costs rise when routes become less dependable. Construction becomes dearer when heavy materials must move through shorter seasonal windows or by air. Insurance terms tighten. Public budgets stretch. Households compensate privately where they can, through bulk buying, deferred travel, or reduced spending elsewhere. These are not merely market fluctuations. They are social signals that infrastructure is becoming less trustworthy. If the state cannot explain or mitigate those shifts, people begin to judge not only the hazard but the hierarchy of concern within public institutions.

Local knowledge often detects this change sooner than formal systems do. Residents know when the roadbed feels wrong under heavy loads, when drainage behaves differently after breakup, when a runway edge softens earlier than it used to, when the usual seasonal sequence no longer holds. Such knowledge is not a substitute for engineering assessment. It is part of the evidentiary field. Where institutions treat it as an anecdote or an inconvenience, they lose both information and legitimacy. Where they integrate it into inspection and maintenance routines, they improve the state's practical sight. The strategic issue is not deference for its own sake. It is whether institutions can recognise emerging instability before the runway defect or road closure forces recognition on harsher terms.

The broader lesson is that roads and runways are not simply infrastructure categories. They are operating conditions for the settlement itself. A community whose routes are unreliable may continue to exist, but under a different political weather: more expensive, more fragile, harder to serve, easier to neglect, and more likely to be described by outsiders as impractical rather than under-supported. This language matters. Once settlements are discussed primarily through cost and difficulty, service withdrawal can begin to sound rational rather than political. Physical fragility then becomes a quiet filter through which decisions about belonging and obligation are made.

For that reason, thaw-related instability should be read not as a remote engineering issue but as a test of whether states can preserve ordinary governability where the ground beneath ordinary systems is changing. The strongest warning sign is rarely the most dramatic failure. It is the moment when local people start planning around the expectation that the road may not hold, the flight may not come, the price may rise again, and the official promise will arrive a season late. At that point, the runway still exists and the settlement still stands. The harder question is whether the state is still present in the full sense of the word, or merely scheduled to land when conditions permit.

Pipelines, storage, and cascading risk

In May 2020, a diesel storage tank at a power plant near Norilsk in northern Siberia collapsed and released fuel into rivers and tundra. The operator was NTEK. The place was the Russian Arctic. The date is easy to verify. The visible event was an industrial spill. The deeper event was a failure of confidence in a whole chain of assumptions about ground stability, inspection, maintenance, and emergency response in a remote environment. The line wavered.

That image of the line matters because pipelines and storage systems are usually imagined as symbols of solidity. They stretch across vast distances, sit inside national energy strategies, anchor local employment, attract capital, and appear in diplomatic language as proof of capability and permanence. Yet linear infrastructure is only as reliable as the conditions through which it passes and the institutions that keep it legible under strain. A pipeline crossing unstable or increasingly variable ground is not merely exposed to physical hazard. It is exposed to inspection burdens, repair delays, response limitations, and reputational shocks that travel far beyond the immediate site of any damage. A storage facility resting on a substrate that behaves less predictably than before does not simply face an engineering challenge. It enters a different political economy of risk.

This is why cascading risk is the right frame. A spill, rupture, or structural failure does not end at the point where fuel escapes or metal deforms. It moves outward through time and organisation. First comes the immediate technical problem: loss of containment, possible service interruption, possible contamination, damage assessment. Then comes access: can inspectors reach the site, can heavy equipment move, are weather and surface conditions narrowing the response window? Then comes prioritisation: which segment is isolated, which traffic or service is interrupted, which community or facility loses supply, which regulator or ministry takes the lead? Then comes the longer wave: liability, insurance, public confidence, investor anxiety, diplomatic scrutiny, redesign, and the quiet rewriting of assumptions about what still counts as safe enough. The line wavers again.

Linear infrastructure is especially vulnerable to this pattern because it joins many local conditions into one long system. Roads, pipelines, and power corridors share a basic difficulty: they pass through terrain that does not fail uniformly. One section may remain stable while another settles, erodes, softens, or shifts under changing freeze-thaw conditions. A single weak point can alter the operating logic of the whole network because lines are only useful when their continuity holds. The concept to name here is distributed exposure. A storage tank can be inspected as a site. A pipeline must be understood as a moving relationship between material, terrain, monitoring, and access across distance. The implication is severe. Strategic fragility often accumulates quietly along segments that appear too ordinary to command sustained political attention until one of them gives way.

The physical stresses are diverse and need no melodrama to become serious. Variable ground conditions can alter support structures, drainage patterns, and alignment. Temperature extremes can change material performance, complicate maintenance timing, and narrow safe working windows. Surface stability may affect whether crews can reach a segment at all. In storage systems, subtle ground movement can undermine assumptions about foundation integrity long before visible failure becomes obvious. None of this requires a dramatic "tipping point" narrative to matter. It requires only that institutions recognise that asset reliability now depends on a more dynamic and less forgiving environment than the one encoded in older maintenance routines.

The Trans-Alaska Pipeline System offers a useful example because it is often invoked as proof that infrastructure can be designed for difficult northern conditions. That is true, and the example deserves respect rather than caricature. The line was built with elevated sections, engineering adaptations, and long-standing maintenance discipline in recognition of seismic and permafrost-related challenges. The relevance here is not that the system disproves infrastructure vulnerability. It shows the opposite in a more disciplined way. It demonstrates that high performance in harsh environments depends on constant attention, specialised expertise, and the acceptance of continuing cost. The lesson is not that linear infrastructure in variable ground is safe by default. It is that safety is an active and expensive achievement.

That counterpoint matters because it prevents the argument from collapsing into fatalism. Pipelines and storage systems do not become ungovernable the moment thaw-related instability enters the picture. Some systems perform well precisely because institutions plan for movement, monitor continuously, and act before visible damage becomes public failure. Yet this limitation sharpens the central claim. Many jurisdictions, firms, and public agencies do not command the same level of technical capacity, financing, or strategic focus. Where maintenance is deferred, inspection regimes are thin, contractor markets are narrow, or response assets are distant, the difference between manageable strain and cascading risk narrows quickly. Good engineering can reduce exposure. It cannot replace governance.

Norilsk made this painfully legible. Whatever the full chain of causation in that specific event, the strategic significance did not lie only in environmental damage. It lay in the way a remote industrial failure triggered broader doubt about substrate reliability, monitoring sufficiency, operator assurances, state oversight, and the practical capacity to respond in time. A spill in a difficult region is never assessed purely as a local accident. Observers read it for what it says about the unseen parts of the system: whether inspection was adequate, whether warning signs were taken seriously, whether remote response capability matched official claims, whether infrastructure assumed environmental stability that no longer held. Confidence in safety regimes can weaken faster than the metal itself.

Inspection is where this often begins. Public debate tends to imagine safety as a function of design and regulation, but linear infrastructure lives or dies through inspection tempo and interpretive judgment. A line can be instrumented, surveyed, and formally compliant while still growing less legible if the conditions around it are changing in ways that exceed existing routines. Remote segments are harder to inspect after severe weather, during seasonal softening, or where access depends on routes with their own vulnerabilities. Storage facilities may present reassuring surface conditions while foundation issues develop more quietly. When environmental variability increases, inspection faces a double burden: more needs to be checked, and some of it becomes harder to check on time. The map is not the site.

Once inspection begins to lag, repair logistics become decisive. A rupture or structural concern does not sit politely in a stable operating environment waiting for a neat intervention. It arrives amid limited airlift, damaged roads, competing emergencies, labour scarcity, procurement constraints, weather shifts, and seasonal deadlines. Heavy equipment may be difficult to mobilise. Spare parts may be far away. Specialist crews may be tied up elsewhere. Remote conditions can turn a technically manageable defect into a prolonged strategic problem because every day of delay expands the cost, the exposure, and the reputational damage. Logistics is therefore not a support function sitting behind safety. It is one of the conditions of safety.

Spill response under remote constraints reveals the same truth with brutal clarity. Public expectations after a release are immediate and often morally charged, as they should be. Yet the practical ability to contain, recover, assess, and communicate depends on whether access routes hold, whether equipment is nearby, whether weather permits deployment, whether local authorities are included early, whether communications function, and whether responsibilities are unambiguous. In remote northern regions, these conditions are rarely generous. Response speed may depend on aircraft that cannot land at full weight, barges that are seasonally constrained, roads that are unreliable, or crews that must be assembled across vast distances. A safety regime that assumes ideal response is not a safety regime. It is a press release.

This is where cascading risk enters finance and diplomacy. Investors, insurers, lenders, and partner states do not examine infrastructure reliability only through direct physical metrics. They also judge the credibility of the institutions surrounding it. Repeated incidents, contested explanations, visible delays in response, or evidence of growing maintenance backlogs can alter perceptions of whether a jurisdiction can manage the assets on which it bases revenue, regional influence, or strategic partnerships. Insurance terms may tighten. Capital may become more expensive. Cross-border projects may face more sceptical scrutiny. Diplomatic trust can fray when one state's infrastructure failures threaten shared environments, shared waters, or shared reputational space. Confidence itself becomes an asset class.

That does not mean every incident produces international crisis. A serious analysis must resist the temptation to inflate all infrastructure failures into geopolitical drama. Many are contained. Some are well managed. Some never travel beyond technical and regulatory circles. Still, the strategic sensitivity remains because pipelines and storage systems sit at the intersection of energy supply, public revenue, environmental liability, and cross-border perception. A line that repeatedly struggles under changing ground conditions may continue to operate, but it may do so within a deteriorating confidence environment. The effect is cumulative rather than theatrical. The cost of capital shifts. Negotiations become harder. Public opposition strengthens. Regulatory tolerance narrows. Strategic room for manoeuvre shrinks.

Temperature extremes complicate the picture further because they stress not only ground conditions but the people and systems that maintain the infrastructure. Heat, cold, smoke, storms, and rapid seasonal transitions affect work windows, material handling, crew safety, inspection intervals, and emergency reach. This matters because safety regimes are often described as though they operate in a neutral administrative climate. They do not. They operate through human and logistical chains that can themselves become more fragile under the same broader environmental conditions affecting the asset. A safety case that ignores the climate stress on its own maintenance apparatus is incomplete.

The politics of reassurance make matters worse. Governments and operators often prefer to speak of safety in binary terms because binary language sounds decisive: the line is safe, the system is secure, the response capacity is in place. But infrastructure under environmental uncertainty rarely deserves binary language. A more honest vocabulary would distinguish between routine operation, heightened monitoring, constrained response, and conditions under which assumptions are being revised. That honesty is difficult because it appears to invite alarm. Yet false reassurance carries its own cost. Once a visible failure contradicts prior confidence, every subsequent claim about inspection, maintenance, and response is judged against the memory of that contradiction. Trust, once cracked, behaves like the line itself.

A second example helps show how this plays out outside dramatic Arctic headlines. Pipeline and storage debates in North America have repeatedly turned on confidence in monitoring, emergency planning, and the ability to respond effectively across long distances and difficult terrain. Public controversy around projects and incidents is not driven only by climate politics or anti-development sentiment, as supporters sometimes claim. Nor is it explained only by corporate negligence, as critics sometimes imply. Often the deeper issue is institutional belief: do affected communities, regulators, investors, and neighbouring jurisdictions believe the safety regime can detect problems early, act quickly in remote conditions, and remain candid when uncertainty intrudes? That question is not rhetorical. It shapes permitting, financing, and diplomatic tone long before any major failure occurs.

This is why local and indigenous knowledge belongs inside the analytical frame even where the immediate subject is linear infrastructure. People who live near these systems often notice changes in drainage, surface movement, timing of breakup, unusual softening, or access difficulties before such patterns are fully reflected in formal reports. That does not make local observation a substitute for engineering data. It makes it part of the evidence base through which safety regimes remain responsive. A governance system that excludes local warning because it arrives in the wrong format may satisfy procedural neatness while losing operational sight. Where trust is already thin, that exclusion also deepens the political cost of later failure.

A fair limitation must be kept in view. Not every pipeline or storage asset in a northern or thaw-sensitive environment faces the same level of risk, and not every shift in ground condition translates into strategic concern. Some systems are overbuilt relative to the stress they face. Some operate with strong redundancy, clear access, and excellent monitoring. Some are old but stable enough to remain manageable. Overstating exposure undermines credibility and can lead analysts to mistake ordinary industrial risk for novel strategic instability. The correction, however, is not complacency. It is discrimination. Which lines face the greatest distributed exposure? Which storage sites sit on assumptions about the substrate that are ageing badly? Which response plans depend on access or staffing conditions no longer reliably available? Those are the proper questions.

The phrase "cascading risk" is sometimes treated as little more than a fashionable synonym for complexity. It should be sharper than that. A cascade has a sequence. In this context the sequence often begins with variable ground or temperature stress, moves into monitoring or maintenance difficulty, then into delayed detection or constrained access, then into service interruption, spill response, liability, and loss of confidence across institutional audiences. One can trace it. Once traced, it becomes possible to see where a state or operator still has leverage in the plain, literal sense of the word: not through rhetoric, but through staffing, logistics, inspection redesign, access planning, public candour, and the willingness to act before visible failure makes every decision more expensive.

The line wavered in Norilsk, but the broader question extends far beyond one tank or one Arctic district. States that depend on pipelines and storage systems in unstable or increasingly variable environments are really making a larger wager: that their safety regimes can remain believable under conditions that are becoming harder to inspect, harder to service, and harder to control at a distance. The wager is not settled by the absence of failure this season, or by a technical report written in the comfort of headquarters. It is settled when the next warning sign appears somewhere remote and inconvenient, and the system must show whether it can still see clearly, move quickly, and convince others that the line has not already begun to waver beyond its own notice.

Maintenance backlogs and procurement tempo

In May 2020, the diesel tank collapse near Norilsk in northern Siberia became a story about pollution, liability, and Arctic fragility. It was all of those things. It was also a story about time. A piece of infrastructure failed in a region where inspection, repair, and emergency response are already shaped by distance, weather, and narrow operating windows. The visible event arrived in a day. The administrative conditions that help explain such failures had accumulated much more slowly: maintenance demands deferred, assumptions left standing, routines built for a steadier substrate than the one now emerging. The season had already shortened.

That is the right place to begin because infrastructure exposure is often described as though it were mainly a matter of design. The roadbed is engineered well or badly. The runway surface is fit or unfit. The storage site is compliant or not. Yet many resilience claims collapse far from the blueprint. They collapse in the procurement office, the vacant technical post, the delayed contract award, the missing spare part, the seasonal barge that arrived after the decision window, the repair crew that could not be booked because three other jurisdictions were already competing for the same expertise. Engineering theory can be sound while the system built to maintain engineering performance is not. The crate stays shut.

This is the argument of tempo mismatch. Environmental change does not wait for annual appropriations, committee review, tendering cycles, staffing approvals, or appeals over contract awards. It alters ground conditions, drainage, freeze-thaw behaviour, and access windows according to physical processes that do not care how public administration sequences its obligations. When those processes accelerate or simply become less predictable, the state may find that its infrastructure is no longer failing according to the same calendar as its repair machinery. A ministry can know what is wrong and still miss the moment at which the fix remained affordable, practical, or sufficient. The damage grows while the paperwork moves.

This is not only a remote-Arctic problem, though remote settings reveal it most starkly. Maintenance backlogs arise whenever assets age under pressure and institutions postpone intervention because present demands are easier to defend than preventive spending. In thaw-sensitive regions, however, backlog behaves differently. A delayed culvert replacement, resurfacing job, embankment stabilisation, or foundation intervention is not simply one more item awaiting funds. It can change the condition of neighbouring assets, narrow the next season's workable repair window, and turn what was once ordinary upkeep into a more complex and expensive operation. The concept to name here is compounded deferral. The implication is severe: the administrative cost of delay rises not only because deterioration continues, but because the conditions under which repair must later occur become less forgiving.

Show the pattern at settlement scale and its politics become plain. A local government identifies runway deformation, road subsidence, or thaw-related drainage failure. Engineers or inspectors recommend intervention. Funding is requested. Procurement rules require competitive process, technical specifications, bid review, perhaps environmental assessment, perhaps specialised contractor qualification. By the time authority is granted and contractors are lined up, the season for heavy works may be ending. The asset is patched rather than repaired. Next year the problem returns, now worse and more expensive. The public sees recurring disruption and hears recurring reassurance. What was described as resilience begins to sound like repetition. Trust then weakens through the visible mismatch between diagnosed need and administrative pace.

Northern transport offers a clear example because it depends so heavily on seasonal timing. Winter roads in parts of Canada have long enabled the movement of fuel, food, building supplies, and heavy equipment. Their strategic value lies not just in cost reduction, but in the planning certainty they historically offered. When cold windows shorten or become more variable, repairs, deliveries, and capital projects must be sequenced more tightly. A delayed procurement decision can mean a whole season lost. A missed season does not simply postpone convenience. It may delay fuel transport, building maintenance, public works, or mine supply in ways that carry into household cost, service quality, and municipal finance. The road does not merely deteriorate physically. It destabilises the planning system that assumed it would behave as before.

Aviation infrastructure in remote regions reveals the same tension with even less margin for error. Many northern communities rely on runways not as supplementary links but as operating necessities. Surface repairs, drainage works, lighting maintenance, and foundation stabilisation must often occur in short windows and with specialised crews or materials. If procurement cycles drag, aircraft restrictions may remain in place longer than expected, heavier loads may not be accepted, and medevac or supply operations may become more vulnerable to weather and scheduling pressure. The problem is not that no one understands the engineering. The problem is that administrative throughput has been calibrated to a slower and more forgiving environment.

This is why maintenance backlogs are strategically more important than they first appear. Public discourse often treats backlog as an accounting category: a deferred amount, a liability on the books, a sign of underinvestment. All of that is true. In thaw-sensitive systems it is also a measure of future fragility. A backlog indicates not only that work remains undone, but that the institution is already behind the environmental tempo of the assets it claims to govern. That lag affects more than the asset. It affects emergency planning, contractor pricing, insurance confidence, and the willingness of local residents to believe official claims that conditions are under control. A backlog is a political forecast disguised as an engineering spreadsheet.

The Norilsk spill remains instructive partly because it condensed these dynamics into one visible failure. Whatever mix of proximate causes one attributes to that event, it drew attention to a wider question: whether infrastructure in changing northern conditions was being maintained and monitored according to the risk actually faced, or according to inherited routines that no longer matched the substrate. That is the broader administrative challenge. Once ground conditions become less reliable, maintenance can no longer be treated as the periodic servicing of a largely stable asset. It becomes an exercise in managing uncertainty under narrowing windows.

Procurement rules are central here because they are where public systems often translate seriousness into delay. Rules exist for good reasons: fairness, transparency, cost control, avoidance of corruption, comparability across bids. These are not trivial goods, especially in infrastructure systems vulnerable to patronage and waste. Yet rules designed for ordinary conditions can perform badly under tempo mismatch. If every intervention must move through procedures calibrated to stable risk and generous timing, then a thaw-related hazard can outpace legality without any law being broken. This is a hard truth for administrative cultures that take process fidelity as proof of competence. One can be impeccably compliant and strategically late.

Contracting capacity creates another pressure point. Remote and northern regions rarely enjoy deep markets of specialised contractors able to mobilise quickly for geotechnical repair, runway rehabilitation, embankment stabilisation, or storage-site reinforcement. The same few firms may be serving multiple jurisdictions, multiple asset classes, and sometimes multiple crises at once. Climatic stress intensifies this scarcity because it can produce concurrent demand across a region. A government may issue a well-formed tender and still find that no suitable bidder can perform within the required window. Or it may secure a contractor only at sharply increased cost because timing has become scarce. The "market solution" then reveals its own fragility: competition works well until the same climate pressure is bidding against itself in several places at once.

Labour follows the same logic. Technical staffing shortages are not glamorous, but they shape resilience more decisively than many high-level strategies do. A missing geotechnical engineer, procurement specialist, airfield maintenance supervisor, or project manager can stall work that funding technically exists to support. Rural and Arctic jurisdictions often struggle to recruit and retain such staff even in calm periods. Under worsening environmental variability, the value of these roles rises precisely when the pool is thinnest. Public systems then confront a rude fact: resilience does not fail only because concrete cracks or gravel shifts. It fails because the person who knows how to specify, authorise, supervise, or inspect the repair was never hired, has already left, or is covering too many responsibilities to move at the pace required.

This is one reason "more resources" is an inadequate account of capacity. Extra money can help, but only if it can be converted into action through staffed offices, available contractors, accessible materials, and decision rights that fit the tempo of risk. A treasury may approve emergency funds after a severe thaw season, yet those funds may arrive too late to salvage the relevant construction window. A ministry may announce an adaptation programme, yet local managers may still wait months for design review or procurement clearance. A state may appear responsive in fiscal terms while remaining operationally slow. The crate stays shut again, this time funded but unopened.

A useful counter-example comes from places where infrastructure systems are maintained through unusually strong institutional continuity rather than episodic reaction. Parts of Alaska's strategic infrastructure, including segments associated with the Trans-Alaska Pipeline System and supporting corridors, have long been managed with the understanding that difficult ground and harsh conditions are not anomalies but standing operating facts. This does not make them invulnerable. It does show that when monitoring, maintenance, and technical staffing are treated as permanent obligations rather than temporary political enthusiasms, some of the worst forms of tempo mismatch can be reduced. The lesson is not that the model can be copied wholesale everywhere. It is that high performance depends on accepting maintenance as a core function of governance, not a budgetary remainder.

That counter-example also marks the limit of any easy prescription. Few municipalities or regional governments possess comparable financial depth, contractor access, or strategic priority. Some assets are simply harder to keep ahead of. Some regions face such thin labour markets and such narrow logistics windows that delay is built into the system from the start. A serious argument should admit this. Not every backlog can be cleared quickly. Not every procurement system can be redesigned without creating new risks of waste or abuse. The point is not to pretend that administrative friction can be abolished. It is to recognise where friction itself has become a material component of infrastructure exposure.

Insurance and finance begin to notice this before politics often does. Repeated temporary repairs, delayed maintenance, and visibly narrowing service windows can alter premiums, exclusions, and lender confidence long before a government formally acknowledges strategic stress. An insurer does not need a grand theory of permafrost change to tighten terms after recurring damage. A municipal creditor does not need to declare a climate emergency to worry about repeated capital deterioration and unstable operating costs. Risk pricing thus becomes an indirect monitor of tempo mismatch. The market is not always wise, and it can misread or overreact. But when it persistently charges for delay, public systems should pay attention. Fiscal signals are often administrative symptoms before they become political arguments.

Local knowledge has an underappreciated role in this domain as well. Residents, local officials, maintenance crews, and indigenous observers often know that a slope is softening, a drainage path is shifting, a road section is degrading earlier in the season, or a runway edge is behaving differently before the formal maintenance schedule catches up. This knowledge is not enough on its own to award contracts or redesign infrastructure. It is enough to challenge the illusion that the current tempo of inspection and repair is adequate. Where institutions ignore such signals because they do not fit standard reporting channels, backlog becomes not only a shortage of work completed but a shortage of hearing. The state falls behind because it has organised itself not to notice in time.

The language of resilience often conceals this entire field of failure. It suggests a quality possessed by systems, a reassuring capacity to absorb shocks and continue. That language is not wrong, but it is too smooth. Under thaw-related exposure, resilience is assembled from procurement calendars, staffing tables, barge schedules, spare parts inventories, inspection rights, seasonal access assumptions, and contractor mobilisation. It lives in warehouses and inboxes as much as in embankments and concrete. When it fails, it usually fails in the translation from known need to completed work. That is why grand claims so often collapse at the level of parts and people rather than at the level of engineering theory.

The strategic point follows directly. Environmental change can outpace administrative throughput without any singular crisis announcing the fact. A government can still repair roads, still fund runways, still commission studies, still hold tenders, and still be losing the race because every intervention lands one season too late. Citizens experience this as recurring disruption and rising cost. Officials experience it as a system forever catching up. Analysts should recognise it for what it is: a structural mismatch between the tempo of hazard and the tempo of governance.

The season had already shortened at Norilsk long before the tank failed. That is what makes tempo mismatch such a useful concept. It directs attention away from the fantasy that resilience is mainly a matter of superior plans and towards the harder question of whether the institutions responsible for roads, runways, depots, and public works can still move quickly enough to keep their promises grounded in the same physical world as the assets they govern. If the environment now changes faster than contracts are awarded, staff retained, parts delivered, and repairs completed, then what exactly does preparedness mean - a funded intention, or a repair finished before the season closes?

Indigenous governance and local knowledge

In 2021, the Yukon government and Indigenous partners in northern Canada continued to confront a familiar problem under increasingly unstable conditions: official monitoring systems could describe part of the landscape, but people living with that landscape often noticed consequential change sooner and in different terms. The place was the Yukon. The period is recent enough to verify. The issue was not romantic insight set against hard science. It was governance. When drainage shifts, ground softens unexpectedly, travel routes become less reliable, and seasonal timings alter, the practical question is not who owns wisdom in the abstract. It is who gets heard soon enough for the state to act before small signs turn into expensive failure. The route is known.

That point matters because local knowledge is still too often treated as cultural decoration in policy writing: something to acknowledge respectfully, perhaps even celebrate, but not something that meaningfully changes thresholds, inspection priorities, procurement decisions, or emergency response. In thaw-sensitive regions, that treatment is not merely patronising. It is operationally foolish. Permafrost instability, shifting drainage, altered freeze-thaw sequences, changing access routes, and subtle changes in surface behaviour often become legible first through repeated attention to place. People who travel a road season after season, land an aircraft on the same strip, maintain a fuel route, or notice recurring changes in ponding, erosion, and ground movement are not offering atmosphere. They are providing observations about system reliability. The route is known again.

This is why local knowledge should be treated as institutional input rather than symbolic consultation. The phrase matters. Institutional input changes decisions. It alters what gets inspected, how often, by whom, with what priority, and under which assumptions. It can redirect scarce maintenance effort before a formal indicator crosses a central threshold. It can refine hazard recognition where remote sensing is patchy, where models are broad, or where general engineering categories miss local variation in soil, water, vegetation, snow cover, and use patterns. The concept to name here is situated detection. The implication is straightforward: a governance system that excludes local observation narrows its own field of vision precisely where physical change is uneven and formal monitoring remains incomplete.

This is not an argument against technical expertise. It is an argument against a false choice that has distorted climate governance for too long. Engineers, surveyors, geotechnical specialists, regulators, and remote sensing analysts do work that local observation cannot replace. They compare, test, model, and document in ways that matter enormously. But they do not stand outside place. Their work is better when it is informed by repeated, grounded, and often community-held knowledge about how a route usually behaves, how breakup has changed, where water now collects, when a shoulder softens, which area has become difficult to cross, and what counts as unusual rather than merely rough. Co-management structures matter because they can turn this relationship into routine rather than exception.

A good co-management arrangement does not simply invite Indigenous communities into a discussion already framed and largely settled elsewhere. It changes the architecture of decision. It creates channels through which local observations can trigger inspection, challenge baseline assumptions, revise maintenance priorities, and influence emergency planning before the state is embarrassed by visible failure. It also does something subtler and equally important: it distributes responsibility more honestly. When hazard recognition depends on both formal monitoring and community-based observation, the fiction that central agencies alone possess the relevant picture becomes harder to maintain. Governance becomes more credible because it becomes more complete.

The value of this is easy to underestimate until one looks at the type of hazards in question. Permafrost-related infrastructure instability rarely presents itself as a single dramatic event at the outset. More often it appears in altered drainage, unusual rutting, slumping, changing snow behaviour, softer shoulders, earlier thaw, or shifts in the timing and quality of access. A satellite may show broad deformation over time. An engineer may record measurable movement once inspection is ordered. A local resident or land user may be the first to say, with useful specificity, that a familiar route no longer behaves as it did two years ago, or that a low section now ponds in a way it did not before, or that the shoulder near a culvert has become unreliable after a certain stage of thaw. These are not anecdotes to be filed politely and forgotten. They are early warnings in human form.

A named example from Alaska helps make the point. In several parts of rural Alaska, Indigenous communities have been deeply involved in observing coastal erosion, unstable ground, changing access routes, and infrastructure vulnerability over long periods. Newtok is especially instructive, not because it offers a simple model, but because it shows the stakes when local knowledge, relocation pressure, and institutional authority do not align smoothly. Community members had long identified the seriousness of erosion and ground instability affecting the settlement. The broader public eventually came to recognise the case as one of the clearest examples of climate-linked relocation pressure in the United States. The lesson here is not merely that local people were right earlier. It is that institutional response lagged despite the presence of local knowledge because knowledge without authority, funding pathways, and decision rights does not move infrastructure on its own.

That lag reveals why legitimacy and effectiveness so often rise or fall together. Public institutions sometimes behave as though legitimacy were a reputational extra, relevant to political tone but secondary to technical delivery. In reality legitimacy is often a condition of technical delivery. A road inspection regime works better when communities believe reporting defects will change outcomes. Emergency response works better when evacuation advice is trusted because earlier concerns were not dismissed. Monitoring programmes work better when local participants see that their contribution is not extractive theatre but part of a decision chain with visible consequences. Legitimacy, in this setting, is visible behaviour: whether reports are answered, whether meetings change priorities, whether data are shared back, whether commitments are kept on the timeline promised. It is not a civic mood floating above the work.

Consultation fatigue enters precisely where that behaviour fails. Northern and Indigenous communities are frequently asked for input on infrastructure, environmental change, relocation, energy development, emergency planning, and monitoring design. The requests can be endless. Maps are marked. Workshops are held. Concerns are recorded. Then little changes, or change arrives in forms that bear only faint resemblance to what communities were told the consultation was for. Fatigue is therefore not simple weariness. It is a rational assessment that participation may be consuming scarce local time without producing proportionate influence. For the state, this is more than a moral failure. It is a degradation of future capability. Once communities conclude that engagement is performative, the quality and candour of future input decline. A governance system can lose sight not because sensors failed, but because people stopped believing the route from observation to action was real.

Data ownership sharpens this problem. Monitoring partnerships often generate valuable information about land, travel, wildlife, infrastructure condition, and local environmental change. Yet the institutional handling of that data can become a point of conflict. Who stores it? Who interprets it? Who can share it? Who profits from it? Who decides what counts as sensitive? These questions are not administrative trivia. They shape whether communities trust monitoring systems enough to participate fully, whether local observations are stripped of context once entered into central databases, and whether institutions can build a durable common picture of hazard. If data flow one way - extracted upward, rarely returned, and only selectively shared - then co-management rhetoric masks an older hierarchy in digital form.

This matters strategically because permafrost and infrastructure exposure are not merely environmental questions. They sit inside land claims, service provision, relocation debates, public spending, and sometimes resource development decisions with implications far beyond a single settlement. A state that treats Indigenous knowledge as supplementary can easily miss blind spots in hazard recognition. A state that treats Indigenous governance as politically inconvenient rather than operationally necessary may preserve central authority on paper while degrading its own ability to govern effectively in practice. In regions where state presence is often judged by the reliability of roads, runways, schools, clinics, policing, and fuel systems, such degradation is not theoretical. It is felt locally and interpreted politically.

Co-management structures can reduce these blind spots when they are real. That requires more than representation at the table. It requires defined authority over monitoring design, clear triggers for inspection or escalation, shared interpretation of local observations, and institutional pathways through which community concerns can affect spending and response. Where these features exist, local knowledge ceases to be an ornamental appendix to expert procedure. It becomes part of the state's sensing apparatus. That matters because formal systems are rarely dense enough on their own to capture all relevant variation across thaw-sensitive terrain. A community noticing that a runway shoulder is softening earlier, that a road embankment is holding water differently, or that a storage site now shows recurrent surface irregularity may supply the difference between preventive action and another delayed repair.

A second example comes from northern Canada, where co-management arrangements connected to land claim settlements and regional governance have, in some places, provided structured roles for Indigenous authorities in wildlife, land use, and environmental assessment. These arrangements are not identical across the North, and their performance varies. Still, they show that local governance can be embedded institutionally rather than appended rhetorically. The relevance to infrastructure exposure is plain. Where such structures exist and are taken seriously, they can provide channels through which changing ground conditions, route reliability, and settlement risks are discussed earlier and with greater place-specific precision. Where they are bypassed or underpowered, central agencies often rediscover local realities only after cost and hazard have already increased.

Still, caution is necessary. Local knowledge is not infallible, homogeneous, or free from internal disagreement. Communities contain different experiences, priorities, and interpretations. What one observer treats as unusual another may regard as normal variation. Some hazards are not easily visible without specialised measurement. Some local reports may be shaped by understandable political stakes around development, compensation, or service demands. To say that local knowledge should be treated as institutional input is not to canonise it beyond scrutiny. It is to insist that scrutiny be reciprocal. Formal models and central datasets are also partial, assumption-laden, and shaped by institutional incentives. The right comparison is not between perfect expertise and imperfect observation. It is between multiple imperfect forms of knowledge that become stronger when arranged honestly.

That reciprocity is often missing because authority remains uneven. A ministry may gladly cite community knowledge in a report while retaining all decisive control over thresholds, budgets, and final interpretation. A regulator may open consultation while keeping the underlying monitoring categories fixed. A company may support community observation programmes while treating uncomfortable findings as context rather than evidence. In each case the appearance of inclusion coexists with a more stubborn reality: the route from knowledge to decision remains centrally owned. This is why arguments about governance matter more than gestures about respect. Respect without institutional consequence is often just a softer form of dismissal.

The deeper reason legitimacy and effectiveness rise or fall together is that both depend on whether people can see their knowledge reflected in action. A community that repeatedly reports changing road behaviour and then receives earlier inspection, temporary restrictions, drainage work, or revised emergency planning learns that the state can listen in operational time. A community that reports the same changes for years while watching studies accumulate and repairs slip learns a different lesson. That lesson spreads. It affects willingness to participate in later monitoring, to comply with warnings, to accept official reassurances, and to believe that future consultation is worth the cost. Legitimacy, once reduced to performance without reciprocity, becomes brittle.

This brittleness is strategically important in remote and northern regions because public authority is already mediated by distance, uneven service, and long histories of imposed planning. Infrastructure instability under thaw conditions magnifies those histories. If the state arrives late to physical change and also treats local warning as secondary, then every subsequent intervention is read against a deeper background of extraction, disregard, or procedural exhaustion. The issue is no longer whether a culvert failed or a runway settled. It is whether the institutions claiming responsibility for the territory are capable of governing it with enough humility and responsiveness to remain believable.

A fair counter-example should be kept in view. Not every co-management arrangement produces better outcomes. Some become procedural thickets. Some are underfunded. Some distribute responsibility so diffusely that no one can act quickly. Some invite participation without giving communities the staffing or technical support needed to engage on equal terms. In urgent settings, layered governance can also slow action if authority is unclear. These are real problems. They show that participation alone is not a cure. But the proper conclusion is not to retreat into central command. It is to design governance arrangements in which local knowledge has defined institutional weight, technical support is adequate, and urgency procedures are clear before crisis tests them.

The route is known in many of these places long before the failure becomes visible to outsiders. People know which roads are behaving differently, which travel seasons have shifted, which buildings are sinking, which shorelines are changing, which airstrips feel less dependable, which repair promises never seem to catch up with the last season's damage. The question is whether a state can build itself to hear that knowledge not as colour, not as grievance alone, but as strategic intelligence about the reliability of the systems through which sovereignty is actually lived. If it cannot, then no amount of formal monitoring will fully close the gap. The route may be known locally and still remain institutionally lost.

Insurance, liability, and fiscal spillovers

In May 2020, the diesel tank collapse near Norilsk in northern Siberia was immediately recognised as an environmental and industrial emergency. The operator was NTEK. The place was the Russian Arctic. The date is easy to verify. Yet the event was also financial from the moment the fuel escaped. Liability attached. Cleanup costs mounted. Confidence in asset integrity weakened. Investors, insurers, regulators, and state authorities were forced to translate a physical failure into monetary consequences. The bill travelled.

That movement from ground instability to fiscal exposure is the real subject here. Infrastructure risk rarely remains inside engineering. It migrates. A runway deforms and repair costs rise. A road softens repeatedly and freight premiums follow. A storage site fails and liability expands beyond the immediate asset. A municipality absorbs repeated damage and its borrowing conditions may worsen, its insurance terms may tighten, or its dependence on higher-level support may deepen. Political systems often prefer to discuss such developments as exceptional expenditures or local financial strain. They are more than that. They are early warnings that the state is losing the cheap fiction that physical reliability and fiscal stability can be managed separately. The bill travelled again.

This is why insurance deserves a central place in any serious account of infrastructure exposure. Insurance is often treated as a technical after-market, a domain for actuaries and lawyers that enters political argument only when homeowners or firms complain about premiums. In reality, it is one of the most disciplined translators of environmental risk into institutional language. Insurers ask whether repeated damage is becoming more likely, whether repair cost is becoming harder to predict, whether access constraints will lengthen claims, whether standards are being updated, whether mitigation is credible, and whether losses can still be pooled on terms that make commercial sense. Those questions may sound private. Their consequences are public. When premiums rise, cover narrows, exclusions expand, or insurers retreat, governments inherit political choices they had hoped to postpone.

The mechanism is simple but unforgiving. A hazard damages or threatens infrastructure. Repeated or severe incidents alter the judgment of insurability. Insurers reprice, impose conditions, narrow coverage, or leave the risk class altogether. Owners then face higher direct costs, uncovered loss, or pressure to self-insure. Municipalities and regional governments may step in because the infrastructure is essential, the community is politically sensitive, or service withdrawal is unacceptable. Public backstops expand, formally or informally. The risk has moved from terrain into the fiscal architecture of the state. The concept to name here is fiscal migration. The implication is stark: governments may deny or soften the climate language around infrastructure risk, yet the budget will often name the problem more bluntly than ministers do.

This is especially important in thaw-sensitive regions because the loss pattern is often attritional rather than spectacular. Repeated subsidence, drainage failure, road repair, foundation work, runway resurfacing, and emergency stabilisation can each be handled as manageable line items for a while. The strategic significance emerges when they start to recur often enough, or in enough combination, that insurance and credit markets read them not as isolated maintenance events but as signs of a deteriorating risk environment. At that point, the fiscal system begins to anticipate what politics may still be reluctant to say: the asset base is becoming more expensive to defend under existing assumptions.

Municipal finance reveals this with particular clarity. Local governments sit close to infrastructure damage and often far from fiscal abundance. They are expected to keep roads passable, water systems functioning, public buildings usable, and emergency access intact, yet they do so within narrow revenue bases and borrowing constraints. When repeated damage events occur, even without dramatic catastrophe, local balance sheets can come under pressure through higher maintenance costs, insurance renewals, emergency works, litigation exposure, and deferred capital expenditure. The effect is cumulative. A town or regional authority may still appear solvent while quietly sacrificing other services, delaying preventive investment, or taking on more contingent dependence on state or federal support. Credit pressure then becomes a weather report of a different kind.

A useful example can be found in repeated floods and infrastructure damage in parts of local government systems across North America, where smaller municipalities have faced pressure not only from the cost of repair but from the financing of resilience work and the uncertainty surrounding future loss. The exact local details vary, and prudence is needed before drawing one-to-one comparisons with permafrost exposure. The common mechanism, however, is clear. Repeated damage alters how lenders, insurers, and public auditors view a jurisdiction's capacity to maintain essential assets without extraordinary assistance. The road does not need to disappear for the fiscal signal to change. It only needs to become expensive and unreliable often enough.

This is where liability broadens the story. Liability is not limited to court judgments after dramatic failure. It includes the wide field of responsibility that emerges when public or private actors are expected to maintain safe systems under changing environmental conditions. A storage operator may face direct liability after a spill. A government may face political or legal pressure if known risks were ignored in public infrastructure planning. A municipality may bear the cost of emergency accommodation, transport, or service continuity when privately insured systems fail to cover the real scope of disruption. Liability is therefore both legal and administrative. It tells institutions how much of the future they may be forced to pay for even when the formal fault line remains disputed.

Norilsk makes this visible because the financial afterlife of the spill was inseparable from its political meaning. What mattered was not only the cost of remediation or the legal assignment of blame, but the broader question of what the event said about infrastructure risk in an Arctic setting where ground stability could no longer be treated as fixed. Once such a question enters the public domain, insurers, lenders, regulators, and diplomatic observers do not simply wait for the next official statement. They start recalculating the risk environment. Confidence becomes more expensive. So does denial.

Confidence is the right word here because safety regimes operate partly through belief. Investors invest not only in physical assets but in the systems that monitor, maintain, and govern them. Insurers price not only expected damage but the credibility of management and response. Neighbouring states and affected communities assess not only direct exposure but also whether institutions will act candidly and competently when things go wrong. Confidence in a safety regime, therefore, has strategic value. It affects capital costs, project timing, international credibility, and the room governments possess to negotiate, reassure, or defer politically awkward decisions. Once confidence weakens, even technically manageable risks may become fiscally and diplomatically larger than their physical footprint would suggest.

Public backstops then enter, often reluctantly. Governments may create explicit support schemes, emergency funds, guarantees, or reconstruction programmes. More often they operate through tacit backstops: insurers retreat but the state cannot allow abandonment; municipal losses mount but the central government cannot let a critical airstrip or access road fail; a community becomes effectively uninsurable yet politically impossible to desert. Such backstops are rational in many circumstances. They can preserve continuity and prevent injustice. But they also reveal something uncomfortable. Risk that private markets no longer wish to carry does not disappear. It is shifted upward, outward, and often quietly onto the public balance sheet. The bill travels once more.

This is one reason fiscal systems can serve as early warning indicators of strategic stress. They record pressure before states always speak about it openly. Premium increases, deductibles, exclusions, reserve drawdowns, rising maintenance appropriations, frequent supplemental budgets, downgraded capital plans, and tighter borrowing conditions all tell a story about how institutions are already absorbing environmental instability. These signs are not perfect. Insurance markets can misprice, overreact, or withdraw too bluntly. Credit conditions can reflect broader economic weakness as well as infrastructure risk. Still, taken together, they often reveal when a jurisdiction is entering a more expensive governing condition before the corresponding strategic language has caught up.

A second example lies in the growing difficulty some jurisdictions have faced in maintaining affordable or available insurance in hazard-prone areas after repeated damaging events. The specific pattern differs between wildfire zones, floodplains, and thaw-sensitive regions, and the analogy should not be pushed carelessly. Yet the underlying logic is consistent. As repeated losses or heightened uncertainty make private cover harder to sustain, governments face a choice: allow retrenchment, subsidise continuity, or redesign the underlying risk environment through more expensive adaptation and regulation. None of these choices is neutral. Each allocates cost, responsibility, and political blame differently. Risk pricing therefore becomes a forcing mechanism. Even when leaders avoid explicit climate language, they must still decide who pays for the changing terms of physical reliability.

This can be seen especially clearly in municipal infrastructure. Roads, schools, clinics, water systems, and airstrips are not optional amenities in remote or northern settlements. They are conditions of ordinary public life. If repeated damage drives up the cost of maintaining them, public authorities may have to reallocate budgets away from other services, seek extraordinary transfers, or postpone work elsewhere. Higher-level governments may treat these as temporary pressures. Local communities often know better. They experience them as a new normal of thinner margins, more emergency procurement, higher service costs, and recurring uncertainty over whether the next repair will be funded in time. Fiscal spillover, in that sense, is political education. It teaches communities how much the state values continuity when continuity becomes expensive.

The phrase "risk pricing" can sound bloodless, but it names a deeply political act. To price risk is to decide, implicitly or explicitly, which losses remain tolerable, which places remain governable on existing terms, and which costs will be socialised. Insurance does this through premium, exclusion, and withdrawal. Governments do it through grants, guarantees, contingent liabilities, and quiet acceptance of deficits or deferred maintenance. Where climate language is contested, risk pricing often tells the truth first. A minister may avoid speaking of thaw-related strategic stress. An insurer tightening conditions on a public asset says, in practice, that the old assumption of stability no longer commands the same confidence.

Still, caution is necessary. Insurance and credit markets are not oracles. They can misread local adaptation capacity, punish communities already disadvantaged, and amplify inequality by pricing bluntly across diverse conditions. Public backstops can create moral hazard if designed badly, preserving exposure without improving resilience. Liability regimes can over- or under-shoot, making some actors absorb costs they cannot manage while allowing others to externalise too much risk. These are real limitations, and they matter. The point is not to treat financial signals as morally authoritative. It is to recognise that they are institutionally consequential. States ignore them at their peril, because they shape the room within which infrastructure policy can still pretend to be merely technical.

The strategic dimension becomes sharper when one looks beyond the single asset. A municipality facing repeated road and drainage failures may endure rising maintenance costs. If insurance terms worsen, borrowing becomes more difficult, and emergency support remains piecemeal, the effects spread. Contractors may demand higher prices. Residents may see taxes rise or services thin. Employers may hesitate to invest. Households may face higher private insurance costs or reduced cover. Migration patterns may shift subtly as those who can move do so. The local state becomes more fragile not because sovereignty has been formally challenged, but because the financial conditions under which it reproduces ordinary order have deteriorated. Strategic harm can arrive by invoice.

This is why infrastructure finance should be read as a form of intelligence. Not perfect intelligence, and never morally innocent, but intelligence nonetheless. It shows where claims of resilience are not convincing those asked to price continuation. It reveals where repeated damage is no longer being treated as bad luck. It exposes the public sectors already functioning as insurers of last resort. It draws attention to communities whose viability is being defended by fiscal improvisation rather than durable design. Used carefully, these signals can help governments recognise which infrastructure exposures are moving from manageable strain into strategic liability.

The administrative response to such signals often determines whether the fiscal system becomes a tool of foresight or merely a record of denial. A government can treat rising premiums, exclusions, and municipal credit strain as unfortunate market behaviour and continue patching assets while urging patience. Or it can treat them as indicators that maintenance regimes, design standards, public backstops, and settlement strategies are no longer aligned with the risk environment. The difference is profound. In the first case, public money follows damage and calls it prudence. In the second, fiscal pressure is read as evidence that the state's underlying assumptions need revision.

Trust also lives inside this financial dimension. Citizens may not read insurance filings or municipal debt disclosures, but they notice when claims are denied, premiums rise, public repairs drag, or local taxes increase after repeated damage. Communities learn quickly whether the state will stand behind essential systems or speak the language of responsibility while shifting more of the cost downward. Trust, in this setting, is visible when authorities explain why spending must change, when backstops are designed before crisis rather than after outrage, when repeated losses lead to altered standards rather than merely larger emergency appropriations. Fiscal credibility is a security asset no less than physical infrastructure is.

The bill travelled from Norilsk into courts, cleanup operations, and public reputations. It also travelled into the larger question this section cannot avoid. When infrastructure built on unstable ground becomes harder to insure, dearer to maintain, and riskier to finance, what exactly is a government protecting when it continues to defend the old settlement pattern or asset design: public safety, political continuity, or the hope that the next premium notice will not say plainly what the last few damage events already have?

Chapter 5

Coastal and Offshore Systems under Warming

What does access mean when a route opens more often yet becomes harder to trust? That question captures the coastal and offshore problem more accurately than the old language of opportunity and obstruction. Warmer conditions and changing ice regimes are often described as though they simply remove barriers, turning difficult waters into more available space. But states, firms, and communities do not plan around abstract access. They plan around reliable access: windows that can be forecast with enough confidence to move fuel, stock depots, rotate crews, schedule repairs, insure voyages, and promise that if trouble starts, help can still arrive before the margin closes. The chart is marked.

That is why coastal and offshore systems should be read less through the romance of opening frontiers and more through the discipline of operating conditions. A port is not merely an asset at the water's edge. It is a timing device for whole regional economies. An offshore platform is not simply a structure at sea. It is a test of how far capability can be stretched before distance itself becomes a hazard. A subsea cable lies out of sight, yet modern emergency services, finance, administration, and public communication often depend on its continued function more intimately than they depend on the visible symbols of sovereignty that draw more attention. Search and rescue, meanwhile, sits at the point where all these systems reveal what they are worth under strain. The chart is marked again.

The central tension runs through each of these domains: warming may alter physical conditions in ways that appear to increase activity, while simultaneously making reliability harder to price, govern, and guarantee. More traffic or longer operating seasons do not necessarily produce more control. They can instead increase the demand for logistics, forecasting, communications, coordination agreements, and emergency response. A system that functions adequately when activity is sparse and conditions are broadly familiar may become more fragile when movements increase across waters and weather patterns that are less predictable than the optimistic narrative suggests. This is not a contradiction. It is the ordinary form that strategic pressure now takes.

That pressure matters because coastal and offshore systems carry more than commerce. They carry the state's practical claims about competence. A delayed resupply changes fuel planning in communities that cannot easily absorb uncertainty. A medevac delayed by weather, distance, or aircraft availability changes not only a single case outcome but also the credibility of an entire operating model. A communications outage from physical cable disruption or delayed repair changes emergency management, public confidence, and the speed at which institutions can coordinate under stress. Search and rescue does more than save lives. It demonstrates whether public authority remains equal to the geography it governs.

The stakes therefore lie not only in physical exposure but in relational capability. Offshore safety is never produced by equipment alone. It depends on nearby vessels, air support, maintenance standards, interoperable communications, clear authority, and rescue assets that are not already committed elsewhere. Port reliability depends not only on quays and cranes but on ice conditions, shipping schedules, warehousing, fuel buffers, and who bears the cost when expected deliveries slip. Subsea connectivity depends not only on cable design but on route redundancy, repair access, specialist vessels, and the willingness to treat communications as critical infrastructure even when damage does not look dramatic from shore. Under warming conditions, each of these relationships tightens.

This is why the chapter's larger concern is not change as spectacle but change as planning burden. Public debate often swings between triumphal claims that the north is becoming more open and apocalyptic claims that infrastructure can no longer hold. Neither framing is disciplined enough. The harder truth is that changing conditions can make activity more frequent and management more exacting at the same time. That combination forces new judgements about buffer stocks, emergency reach, repair capability, and the degree of slack public systems can still afford to carry. Efficiency begins to look different when the cost of being wrong about timing rises.

The path ahead follows the operating logic of that burden. It begins at the port, where variable access exposes the difference between movement and reliability, then moves outward to the offshore platform, where distance and weather narrow the margin between routine operation and emergency dependence. From there the focus drops beneath the surface to communications links that are easy to ignore until their failure reorganises the whole response environment, before arriving at search and rescue as the clearest public expression of whether capability remains credible when several systems are stressed at once.

What joins these themes is a single strategic fact: coastal and offshore systems are governed through promises about response. A community is told that fuel will arrive within a workable window. A crew is told that evacuation, medevac, or containment support can be mobilised if conditions deteriorate. A government assumes that emergency communications will still function when ordinary links fail. Those promises may be prudent, overconfident, or quietly obsolete. Warming does not judge them morally. It tests them operationally. When the weather shifts, the ice behaves unexpectedly, or a repair vessel cannot reach the cable on the hoped for schedule, the issue is no longer whether a plan existed. It is whether the state arranged enough capability close enough to matter.

That is also why sovereignty enters the story in a form less theatrical than strategic literature often prefers. Presence is not only patrols, flags, and legal claims. It is the repeated demonstration that life, work, and distress in exposed waters remain connected to institutions that can still answer. Ports, platforms, cables, and rescue services therefore belong in the same frame because they are all parts of the state's reply to distance. When that reply slows, confidence weakens first in local and operational circles, then more broadly in finance, politics, and diplomacy.

The harder question waits just offshore. If access can no longer be measured simply by whether the ice retreats or a route appears open, what should count as the real metric of control: the waterway itself, the timetable for response, the resilience of communication, or the ability to keep a promise when several of those fail together?

Ports, sea ice variability, and reliability

In July 2013, the cargo vessel *Nunavik* completed a voyage through the Northwest Passage without icebreaker escort, carrying nickel concentrate from northern Canada toward markets in Europe. The ship belonged to Fednav. The route passed through waters that, for generations, had symbolised obstruction rather than routine transit. The episode was widely read as proof that Arctic access was becoming easier. It was certainly read that way by commentators eager for a dramatic shorthand: less ice, more movement, fewer barriers. Yet planners who live by delivery windows rather than headlines would have drawn a more cautious lesson. A route can open and still be hard to trust. The window matters.

That distinction lies at the heart of coastal and offshore logistics under warming conditions. Sea ice variability alters access, but it does not do so in a simple, linear direction from difficult to easy. Ports, resupply systems, and shipping schedules are not organised around the abstract possibility of movement. They are organised around dependable timing. A community waiting on fuel does not ask whether a ship *could* reach it under ideal conditions. It asks whether the delivery can be expected within a workable range, whether storage can bridge delay, whether substitute transport exists if the window narrows, and who bears the cost when prediction fails. Reliability is the stricter word.

The popular image of a warming Arctic often turns on opening. Ice retreats. Routes appear. Commercial opportunity expands. Some of that is true. Yet changing ice regimes can also produce more unstable planning conditions: thinner or less predictable ice, shifting seasonal transitions, fog, storms, swell in waters once buffered differently, ice movement that complicates navigation, and greater difficulty in knowing whether a route open on paper will remain dependable across the whole supply chain. The concept to name here is variable access. It matters because systems built on narrow seasonal windows do not become secure simply because one form of barrier weakens. They become newly dependent on forecasting, buffer capacity, and risk decisions that are harder to standardise.

Ports feel this first because they are timing machines as much as physical facilities. A port in the north or sub-Arctic is not valuable merely because it can receive ships. It is valuable because ships can be received, unloaded, and turned around in a sequence that supports warehousing, onward transport, fuel planning, and service continuity. If sea ice and weather regimes become more variable, a nominally longer season may still produce greater administrative strain. Deliveries arrive too early for one storage pattern, too late for another, or in less predictable clusters that complicate labour, berth use, and downstream distribution. The issue is not movement in the abstract. It is whether the choreography still holds.

This is why reliability, not openness, should be the planning metric. Reliability means a route performs within a range narrow enough for decisions to be made without expensive overcompensation. Shipping firms need it to sequence crews, manage insurance, and allocate vessels. Port authorities need it to organise berths, warehousing, and repairs.

Communities need it to plan fuel stocks, food deliveries, construction materials, and public services. Governments need it because resupply failure quickly becomes political. If the same route is theoretically accessible more often but practically dependable less often, then warming has not simplified governance. It has made it more exacting. The window matters again.

Northern communities in Canada offer a clear illustration through annual marine resupply. In many Arctic settlements, barges and sealift operations deliver the bulk goods that make ordinary life possible: fuel, building materials, vehicles, household goods, and supplies too expensive to move by air. These deliveries already depend on careful seasonal planning. If ice conditions, weather, or port access become less predictable, communities and suppliers face a harder choice. They can carry larger buffers and accept higher storage and financing costs, or they can operate leaner and risk disruptive shortages if a vessel is delayed or a route becomes unfavourable at the wrong moment. This is not a theoretical dilemma. It is logistics becoming politics.

Fuel planning shows the stakes most sharply. A delayed delivery does not remain a shipping problem for long. It becomes a problem for electricity generation, heating, municipal budgeting, airline schedules, school operation, water systems, and local confidence in public assurances. A government may speak of climate adaptation at a high level while a community asks a plainer question: will the fuel arrive before the margin closes? If the answer becomes harder to predict, then the state must either invest in larger buffers, accept greater fiscal exposure, or live with a higher risk of emergency improvisation. Reliability, in this setting, is a budget line in disguise.

A named example from Alaska helps reveal the same logic from another angle. The annual summer barge season in many western and northern Alaskan communities has long been crucial for bulk fuel and freight. The specific conditions vary across ports and coasts, and it would be careless to collapse them into one Arctic story. Still, the governing challenge is familiar. Marine access depends on a mixture of ice, weather, coastal conditions, infrastructure limits, and scheduling precision. When those variables behave less predictably, the state and suppliers cannot merely celebrate access. They must recalculate storage, contingency transport, insurance, and the political consequences of delay. An open route that cannot be trusted is not the same thing as a secure supply.

The temptation to read all this through a language of opportunity has been strong for years. Governments, firms, and commentators have often treated reduced sea ice as though it automatically lowered the cost of operation and widened the margin for error. That reading contains a half-truth and omits the harder half. Some voyages do become feasible that once were not. Some transit times may shorten under some conditions. Some infrastructure projects become more commercially viable. Yet for planners concerned with ports and communities, the core issue is not whether movement has become possible in isolated cases. It is whether the full chain of movement has become governable enough to rely on. A route successfully used once can still be a poor basis for annual public planning.

This is where logistics buffers enter the argument. A buffer is often described as inefficiency: extra fuel in storage, additional inventory, slack in shipping schedules, spare warehousing, conservative timing. Under stable conditions such slack can indeed look expensive. Under variable access, it becomes a form of insurance purchased in kind rather than a contract. The more uncertain the delivery window, the larger the buffer needed to keep public and private systems from becoming brittle. That increases cost. It also shifts the political argument. Who pays for the additional storage? Who funds expanded tankage, warehousing, or reserve purchases? Who absorbs the loss if goods arrive too late or if extra stock proves unnecessary in a milder season? Once reliability weakens, disputes over responsibility follow naturally.

Those disputes are not trivial administrative quarrels. They can expose deeper tensions over who is expected to carry the cost of environmental uncertainty. Suppliers may argue that governments must subsidise buffers for essential communities. Governments may insist that operators price risk more realistically. Municipal authorities may warn that they cannot finance larger reserves or replacement infrastructure alone. Residents may see each round of blame as proof that state commitment is conditional and slow. Political legitimacy, here, does not rest on speeches about Arctic sovereignty. It rests on whether fuel arrives, shelves stay stocked, and no one has to discover the fragility of the system in the week before freeze-up.

A second named example shows how quickly this becomes strategic. During periods of delayed or constrained resupply in northern communities, public concern often centres not only on price but on whether official systems anticipated the risk. The details vary by year and region, and one should resist turning every late shipment into a grand narrative. But the pattern matters. If deliveries are delayed and fuel planning tightens, local officials may have to alter non-essential use, postpone works, or coordinate emergency measures. What appears to an outside observer as a scheduling problem is experienced inside the community as a test of whether institutions understand how narrow the operating margin really is. A delayed barge can teach more about state capacity than a policy white paper.

This is why sea ice variability should not be described merely as a navigational issue. It changes the economics of trust. Insurers, charterers, and investors care about whether routes can be forecast well enough to underwrite exposure with confidence. Port authorities care about whether dredging, repairs, labour, and berthing decisions can be aligned with vessel movement that no longer follows older patterns. Communities care about whether the promised system remains one around which they can organise life. These are different audiences, but they are all reading the same thing: not the presence or absence of ice alone, but the dependability of a chain of arrangements built upon that environment.

Still, caution is necessary. Not every change in ice regime makes access less reliable. Some ports or routes may genuinely become more workable over certain periods, and some operators may adapt effectively through better forecasting, more flexible scheduling, specialised vessels, and revised logistics practices. A serious argument should admit this. The claim is not that warming makes every maritime system more fragile in every respect. It is that accessibility and reliability are different categories, and the first can improve while the second becomes harder to guarantee. Analysts who conflate the two risk misunderstanding what communities and planners actually need.

That distinction becomes even more important when one remembers that ports are not isolated points. They are nodal assets in longer supply systems. A delayed ship affects storage decisions inland. A missed resupply changes construction sequencing. A fuel delivery problem ripples into electricity generation and aviation. A disrupted port call can force air freight substitution at far higher cost. The maritime issue then spreads into municipal finance, household budgets, and public service continuity. Reliability matters precisely because so many other systems silently depend on it. Once the port window becomes less predictable, the burden is redistributed across the whole territory served by that port.

Search and rescue, ice information services, and emergency response capacity sit just behind this logic even when they are not visible in the shipping schedule. A route is more governable when mariners know what support exists if conditions deteriorate. It is less governable when rescue distance is long, communications are patchy, and weather can outpace the assumptions built into voyage planning. Capability, in this sense, is relational. A port is safer not only because of its own infrastructure but because of the wider network of services that make arrival and departure less final a gamble. If activity increases while that surrounding capability does not, the apparent gain in access may conceal a tighter operational margin.

The phrase "opening Arctic" has survived partly because it is vivid and easy to repeat. It has also survived because it flatters strategic fantasies about mobility and commerce. Yet for the communities and institutions that rely on marine resupply, the more honest phrase would be something like "uncertain access under growing dependence". Less memorable, perhaps. Much closer to the governing problem. What matters is not the symbolic breakthrough of one voyage or one season, but whether the timetable remains stable enough to anchor fuel planning, food delivery, and public commitments without increasingly expensive buffers and recurring political conflict.

The window mattered for *Nunavik* in 2013, and it matters even more for the less glamorous traffic that keeps northern life functioning. A cargo vessel proving passage can inspire headlines. A delayed fuel delivery can rewrite a community's winter. That is why reliability belongs at the centre of analysis. If changing ice regimes widen the possibility of movement while narrowing the certainty on which planning depends, then the key question is no longer whether the water is open. It is who is expected to pay, store, wait, and apologise when the window shifts after the promise has already been made.

Offshore platforms and emergency response distance

On 20 April 2010, the Deepwater Horizon drilling rig exploded in the Gulf of Mexico. The operator was BP. The event has entered public memory chiefly through fire, oil, and the scale of the spill that followed. Yet one of its most enduring lessons lies in something less visual and more operational: distance. Between platform and shore, between alarm and assistance, between specialist equipment and the moment it is needed, there exists a stretch of time that no amount of technical confidence can wish away. The helicopter has to arrive. The vessel has to reach the site. Weather has to permit action. The gap is real.

That gap matters far beyond one disaster. Offshore platforms are often described through engineering sophistication, production capacity, and regulatory oversight. All of these matter. None abolishes geography. A platform may be highly advanced, well equipped, and staffed by experienced crews, yet still exist within an operating environment where response time is the hardest fact. When conditions are stable and support assets nearby, distance can be managed, priced, and folded into ordinary procedure. When weather shifts, visibility falls, sea state worsens, or several demands compete at once, the same distance becomes a strategic constraint. The waiting defines capability.

This is why offshore safety under warming conditions cannot be understood as a simple contest between technology and hazard. It is a matter of operational reach. A platform at sea depends not only on what is installed on the structure itself, but on medevac aircraft, standby vessels, spill response assets, communications links, weather intelligence, crew transfer systems, and the agreements that determine who moves first, who authorises what, and how support is sequenced once trouble begins. The concept to name here is relational capability. The implication is straightforward and severe: an offshore installation is never as safe as its equipment alone suggests. It is only as safe as the network that can reach it under the worst plausible conditions in which that network must still function.

This point is often obscured by the way offshore infrastructure is presented publicly. Governments and operators tend to emphasise design standards, technical systems, and compliance regimes, because those are legible signs of seriousness. They can be audited, certified, and displayed. Distance is harder to stage. It sits in flight time, sea conditions, vessel location, crew fatigue, fuel range, and the availability of specialised support that may already be committed elsewhere. Yet distance is where many safety margins truly live. If a casualty needs evacuation, minutes and weather windows matter more than the elegance of organisational charts. If a release begins, containment depends not only on whether equipment exists somewhere in the system, but on whether it can be deployed before the problem outruns the response.

Extreme weather variability tightens these margins even where advanced equipment is available. This does not require apocalyptic conditions. A narrower flying window, heavier swell, unexpected ice movement, freezing spray, poor visibility, or a run of rough conditions that delays crew rotation and vessel transfer can each turn a theoretically manageable problem into a more dangerous one. Offshore operations are already calibrated to uncertainty. What changes under warming conditions is not simply that the weather worsens in some generic sense, but that familiar assumptions about timing and operating windows become less dependable. A medevac plan can be sound under median conditions and fragile under the specific pattern of volatility that now arrives more often or with less warning than the paperwork assumed.

Show the issue at the level of emergency medicine and the argument sharpens. A worker becomes critically unwell or seriously injured on a platform. Onshore care exists. The platform has trained personnel, initial treatment capacity, communications, and a protocol for escalation. Everything now turns on the bridge between incident and definitive care. Can a helicopter launch? Can it land or winch safely? Is there a vessel closer than the shore base? Is another aircraft already committed? Is the receiving hospital prepared? Does weather permit one route but not another? The relevant metric is not "is there a medevac plan?" but "how many contingencies can be absorbed before the plan ceases to work as imagined?" That is a much harsher question. It is also the right one.

The same logic governs spill containment. Public imagination often pictures response as a matter of equipment stockpiles and specialist teams. Those are necessary. They are not enough. Offshore spill response under remote or variable conditions depends on reach, sequencing, and endurance. Equipment may be stored at one port, specialist vessels at another, aircraft at a third, with legal authority divided among operator, coast guard, environmental agency, and perhaps military or regional actors. In calm water and good visibility, this may be coordinated effectively. Under deteriorating weather, rough seas, ice, or simultaneous demands, the friction multiplies. Response time lengthens. Containment options narrow. Public confidence falls precisely because the existence of capability somewhere in the system is revealed to be different from capability at the site in time.

The Deepwater Horizon case remains instructive not because every offshore platform faces the same risk profile, but because it exposed how quickly technical failure becomes a multi-layered crisis in which response distance shapes consequence. Firefighting, search and rescue, spill control, medical response, investigation, and public communication all had to operate across sea and institutional boundaries. The lesson for this book is not to revisit that disaster as moral spectacle. It is to recognise that offshore systems are never self-contained. Once something goes wrong, capability is tested through moving parts distributed across agencies, contractors, vessels, aircraft, ports, and weather windows. Distance is not a backdrop. It is part of the event.

A second example, less catastrophic but equally revealing, lies in the routine operational logic of northern offshore work in places such as the Norwegian and Barents Sea regions. These environments have long forced operators and regulators to take weather, cold, and reach seriously. The relevance here is not to present any one national system as exemplary beyond criticism. It is to note that safety in these settings is openly built around coordination among nearby assets, standby vessels, helicopter capacity, weather services, and clear rules about readiness and response. That is exactly the point. Capability is not located only on the platform. It is distributed. Where that distribution is dense, drilled, and well governed, margins widen. Where it is sparse, assumed, or administratively fragmented, margins narrow even if the platform itself is modern.

This is why the phrase "advanced equipment" can mislead. Advanced equipment may improve monitoring, shutdown procedures, onboard medical care, firefighting, and communications. It may reduce some forms of risk substantially. But it can also encourage a false narrative of self-sufficiency if policymakers or investors begin to treat technological sophistication as a substitute for proximity and coordination. A platform fitted with excellent systems remains vulnerable if medevac relies on a single aircraft rotation, if spill response assets are too far away, if mutual assistance between installations is uncertain, or if weather variability increasingly interrupts the assumptions behind emergency planning. Technology can reduce hazard exposure. It does not abolish distance.

This becomes especially important as offshore activity grows or shifts into more exposed waters. Increased movement at sea can generate a comforting sense that a region is becoming more active and therefore more serviceable. Sometimes it does. More often activity stretches support systems before public institutions fully admit that they are being stretched. A port may serve more traffic without a proportional increase in rescue coverage. Helicopter availability may remain constant while platform numbers, shipping movements, or tourist voyages grow. A standby vessel may cover a wider area than before because budgets are tight and policymakers assume the most dangerous scenarios remain exceptional. Capability then begins to erode not through neglect of safety language, but through overconfidence in what existing arrangements can absorb.

The word *relational* deserves emphasis because it clarifies where failure often begins. An offshore operator may have strong internal procedures. A coast guard may have serious assets. A health service may have air ambulance capability. A neighbouring installation may have standby support. Each part may function competently on its own terms. The problem arises when the relationships among them are not as settled or as fast as the planning assumption requires. Whose aircraft takes priority? Who authorises cross-jurisdiction deployment? Which communications system is primary if another fails? How quickly can environmental data be shared? What happens if a platform emergency coincides with a maritime rescue or a coastal disaster? The answers to such questions are usually buried in agreements and exercises, not in public rhetoric. Yet those buried answers often decide the outcome.

Mutual aid agreements illustrate this vulnerability perfectly. They are often cited as evidence of preparedness, and sometimes rightly so. They allow operators, agencies, and even neighbouring states to pool scarce assets and share reach. But mutual aid is only as strong as the conditions under which it was imagined. If agreements assume staggered incidents, benign weather deterioration, or spare capacity in nearby systems, they can prove brittle when conditions change. A severe offshore event that occurs during regional storms, high demand on rescue services, or concurrent coastal emergencies can reveal that the same assets have been promised several times over. The paper holds. The aircraft does not arrive. The waiting returns.

This is where warming matters in a more exacting way than public shorthand usually allows. The issue is not merely that sea conditions become "worse". It is that weather variability, changing ice regimes, fog, storm tracks, and other operational conditions can alter the reliability of the support network upon which offshore safety depends. A platform may remain technically serviceable while the surrounding envelope of response becomes more uncertain. That is a strategic shift because it changes the meaning of preparedness. What looked like acceptable distance under one climate pattern may look more exposed under another, even if the platform's hardware has not changed at all.

Search and rescue sits close to this argument even before the later chapter treats it directly, because rescue is one of the clearest public measures of offshore competence. A state or operator can write impressive risk frameworks, but when a rescue is needed, the visible facts become simpler. How long did it take? Under what conditions did assets move?

Were communications clear? Did coordination hold? Were casualties recovered? Did response remain available for the next incident, or was the system temporarily emptied by the first? These are not only humanitarian questions. They shape public belief about whether offshore expansion, infrastructure maintenance, or coastal governance still rest on credible foundations. A rescue operation is also a political demonstration.

Still, caution is necessary. It would be wrong to claim that every increase in offshore distance or every spell of difficult weather produces strategic fragility. Some regions have built impressive systems of redundancy, forecasting, standby support, and interagency coordination. Some operators maintain high standards precisely because they recognise that response distance cannot be wished away. Some weather variability is already built into conservative planning. The point is not that advanced offshore work has become impossible or uniformly reckless. The point is that the margin between ordinary operation and dangerous delay is relational and weather-dependent, and therefore more vulnerable to environmental volatility than celebratory narratives of technology and access tend to admit.

A fair counter-example lies in sectors where rigorous planning and dense support have produced comparatively strong performance in harsh conditions. Parts of the Norwegian offshore regime, for instance, are often noted for integrating operator obligations, rescue capacity, and a strong safety culture more tightly than many jurisdictions do. This matters because it shows that distance can be governed seriously. Yet even here the lesson is not comfort. It is burden. Strong performance requires sustained investment, training, coordination, and political willingness to fund capacities that may appear underused in ordinary periods. Many governments and firms prefer to rely on thinner margins until a visible shock exposes what those margins were worth. The existence of good practice therefore underscores the cost of pretending that distance is only a technical matter.

Public trust follows from these arrangements in ways that are rarely acknowledged directly. Offshore operations depend on workers believing that if something goes wrong, the system behind the platform is real. Coastal communities depend on believing that if a spill occurs or a rescue is needed, the institutions claiming control over the sea can actually reach it. Investors depend on believing that a jurisdiction's safety claims are backed by assets and agreements rather than paper confidence. Trust, then, is visible behaviour: aircraft launched on time, vessels positioned sensibly, communications functioning across agencies, protocols revised after exercises, difficult truths spoken early when weather narrows the response window. Trust is not a sentiment appended to safety. It is one of the things safety consists of.

This is why the politics of offshore development often become arguments about distance even when they are framed as arguments about technology, jobs, or sovereignty. Beneath the headlines lies a plainer question: who can get there, how fast, under what conditions, and with whose authority? If the answers are vague, overconfident, or dependent on too many good assumptions arriving together, then offshore capability is weaker than the equipment inventory suggests. The platform may be impressive. The network supporting it may not be.

The waiting defined Deepwater Horizon, and it continues to define offshore risk wherever platforms depend on help that must travel across difficult water and uncertain weather. That is the unresolved pressure this section leaves in view. When policymakers say an offshore region is ready, what exactly are they claiming readiness about: the installation itself, or the chain of aircraft, vessels, hospitals, ports, agreements, and weather windows that must all function before "ready" means anything at sea?

Subsea cables and communications dependence

In January 2022, the volcanic eruption and tsunami near Tonga severed the country's main undersea communications cable and left the islands largely cut off from the outside world. The place was the South Pacific. The date is easy to verify. The immediate drama belonged to geology and ocean force. The deeper lesson belonged to infrastructure. A modern state had not lost only internet access in some consumer sense. It had lost a large part of its ability to coordinate, reassure, transact, report, and connect emergency need to institutional response. The line went quiet.

That silence is the right place to begin because subsea cables are among the least visible pillars of contemporary state function. They lie out of sight and beneath public imagination, which is one reason they are so often discussed only when sabotage is suspected. Sabotage is one possible risk. It is not the only one, and not always the most analytically useful starting point. Climate stress, geological disruption, coastal erosion, storm impacts on landing sites, delayed repair access, and the sheer operational difficulty of restoring damaged marine infrastructure can all expose the same fact: connectivity is not virtual. It is physical, routed, serviced, and vulnerable to time. The signal has weight.

This matters because digital dependence has spread further than many public institutions still speak as though it has. Emergency communication, telemedicine, air traffic coordination, banking, welfare administration, customs, shipping documentation, weather data exchange, public warning systems, education platforms, and routine government communications all rely, directly or indirectly, on resilient data links. When those links are interrupted, the resulting disruption is rarely confined to one ministry or sector. A cable fault can begin as a technical issue and become a governance issue within hours because modern systems are layered upon the assumption that information will continue to move. The concept to name here is connective dependency. The implication is severe: the more silently a society relies on data continuity, the more dangerous it becomes to treat cable resilience as a specialist concern rather than a security concern.

That danger is often obscured by the language used around communications infrastructure. Officials speak of bandwidth, resilience, capacity, redundancy, and outage management. Each term has technical meaning. None captures the whole strategic picture. A state does not merely need data to flow under ordinary conditions. It needs enough redundancy, repair access, and institutional clarity that communication remains usable when environmental disruption has already stressed transport, energy, or emergency services. A single subsea route may look efficient and commercially sensible until a fault occurs during a storm season, a volcanic event, a period of rough sea, or some other condition that delays repair vessels and limits alternative routing. Then efficiency is revealed as concentration. The line goes quiet again, and this time many other systems begin to speak its absence.

The Tonga case remains especially instructive because it stripped away the comforting fiction that connectivity is somehow too distributed to fail cleanly. When the cable was damaged, the consequences reached well beyond inconvenience. International communication slowed drastically. Families struggled to reach one another. Officials had reduced ability to communicate externally. The flow of information needed for emergency assessment and coordination narrowed sharply. The issue was not only how to restore internet traffic. It was how a geographically dispersed island state would function administratively and socially while one of its principal channels to the wider world was impaired. A cable is often described as infrastructure. In such moments it becomes constitutional in feel.

A second example comes from the Shetland Islands in October 2022, when subsea cable damage contributed to communications disruption that affected phone services and broadband. The incident did not produce the same global shock as Tonga, but it was strategically revealing in a different way. Here was a highly developed setting in which damage to undersea infrastructure quickly raised concerns not only about everyday connectivity but about the continuity of emergency calls, institutional coordination, and the dependence of routine public life on systems that most users rarely think about until they fail. The significance lies not in the uniqueness of the case, but in its ordinariness. One need not imagine extreme catastrophe to see how physical damage below the waterline can rearrange the practical capacities of a community.

Physical damage is central to this argument because it restores a sense of materiality to a domain often misdescribed as weightless. Subsea cables can be affected by seismic activity, volcanic events, fishing gear, anchors, seabed movement, coastal erosion at landing points, and difficult marine conditions that complicate both prevention and repair. Climate stress enters this field less as a single mechanism than as a worsening context. More volatile coastal conditions can threaten landing stations and nearshore environments. Stronger or less predictable marine conditions can complicate maintenance and repair access. Concurrent coastal hazards can place strain on the same transport and emergency systems that would otherwise support restoration. The point is not that climate alone explains cable damage. It is that climate stress can tighten the operational margin within which cable resilience is managed.

This is why repair access matters almost as much as route design. Public discourse often focuses on whether a cable exists and whether a backup route also exists. Those are important questions. They are not enough. Once a fault occurs, everything turns on whether specialised repair vessels can be mobilised, whether weather and sea conditions permit work, whether permits and coordination are in place, whether spare cable and technical teams are available, and how long the affected system can operate in degraded mode before knock-on effects begin to multiply. A cable is not resilient because someone once paid to lay it. It is resilient when the entire chain of detection, prioritisation, access, and restoration has been thought through under difficult conditions.

This is where redundancy moves from technical preference to security principle. Redundancy is often treated as expensive duplication until failure makes it look like foresight. A second route, a satellite backup, a hardened landing station, or an alternative communications protocol may all appear underused in calm periods. Under stress they can preserve emergency communications, financial continuity, and public messaging long enough to prevent one infrastructure failure from triggering several institutional ones. The important point, however, is that redundancy must be judged in relation to real operating conditions rather than procurement optimism. A nominally separate route may share exposure at the landing point. A satellite fallback may offer only limited capacity. A backup system may not support the services that matter most during emergency response. Redundancy on paper is not the same thing as redundancy that holds.

Emergency communications reveal this distinction immediately. When a subsea cable fails, public institutions do not simply lose speed. They may lose hierarchy. Dispatch systems can slow. Data-dependent emergency coordination can become patchier. Hospitals may lose access to systems needed for referral, consultation, or logistics. Police, coast guard, fire, and civil authorities may find that their interoperability depends more on ordinary telecommunications than planning documents had admitted. The knock-on effects are rarely cinematic. They are administrative and cumulative. Response times lengthen. Information arrives unevenly. Public warnings become harder to update. Rumour fills the silence. Capability is reduced not because the institutions have forgotten their duties, but because the network binding them has thinned at the very moment when stress demands tighter coordination.

The reader should resist one common mistake here. Not every cable outage is a strategic emergency. Some are brief, contained, and effectively managed through rerouting or backup systems. Some jurisdictions have enough redundancy to absorb faults with little public consequence. Some repair ecosystems are strong enough that restoration proceeds before serious secondary effects take hold. These are real limitations on the argument, and they matter. To speak of communications dependence as a security issue is not to inflate every technical fault into national crisis. It is to insist that the strategic importance of a cable lies not in the drama of failure but in the degree to which essential public functions rely upon it without adequate alternatives.

That limitation actually strengthens the analysis because it directs attention to variability in state performance rather than to universal alarm. A cable fault in one place becomes an inconvenience. In another it becomes a test of governmental continuity. The difference lies in route diversity, backup planning, landing-site resilience, regulatory clarity, repair access, emergency protocol design, and the habits of interagency coordination. A wealthy state with multiple routes and strong institutional preparation may absorb an outage with tolerable friction. A remote island or sparsely connected region may find that the same class of damage reaches deep into emergency response, finance, and public trust. The physical event may be similar. The political consequence is not.

This also explains why subsea cables belong inside the same frame as ports and offshore platforms, even though they carry information rather than fuel or cargo. All three are systems of reach. All three depend on marine access, repair logistics, and supportive infrastructure that may sit far from the point of immediate use. All three are vulnerable to weather, distance, and the difference between capability somewhere in the system and capability at the site in time. And all three expose a broader truth about contemporary governance: infrastructure once considered supportive has become constitutive. Communications are no longer adjuncts to state function. They are among its operating conditions.

A further complication lies at the landing site, where the apparent boundary between sea and land can disguise a dense zone of risk. Landing stations, coastal conduits, power supply, backup systems, and local access roads may all be more exposed than national narratives of digital strength suggest. Coastal erosion, flooding, storm surge, and power instability can turn a subsea cable issue into a terrestrial infrastructure issue with surprising speed. The strategic lesson is simple. Connectivity does not fail only on the open seabed. It can fail at the mundane edge where marine infrastructure meets neglected local systems. That is where climate stress often enters most plainly, and where responsibility is most likely to fragment across agencies and owners.

Trust, often invoked too vaguely, takes on concrete form in this domain. A population trusts communications infrastructure when emergency calls connect, when public warnings continue through disruption, when banks and basic transactions still function, when local authorities can communicate with national authorities, and when officials explain outages and repair timelines candidly enough that uncertainty does not become suspicion. Trust degrades when people discover that a supposedly resilient system has one effective route, that backup systems support only some users, or that restoration depends on vessels and permissions that were never integrated into serious planning. Trust, here, is not a sentiment added after the engineering. It is one of the products of engineering and governance done well.

This is also where interdependence begins to look strategic in the strongest sense used in this book. A cable fault can complicate emergency services. Emergency services under strain may then struggle to support repair logistics. If ports or airports are also affected by weather or coastal disruption, the timeline stretches further. Public communication grows thinner just as the need for clarity increases. Economic activity slows. Political pressure rises. None of this requires sabotage, and none of it requires a spectacularly weak state. It requires only a sufficiently concentrated dependence on connectivity and a sufficiently narrow margin for repair. Single points of failure are often invisible until the system begins to route around them and cannot.

A fair counter-example deserves notice. Some states and regions have invested heavily in route diversity, satellite resilience, cable landing security, and emergency communications fallbacks. In such places, cable damage may produce manageable degradation rather than strategic stress. This is worth emphasising because it prevents the argument from sliding into determinism. Connectivity dependence is not destiny. It is a planning condition. But the counter-example also clarifies the burden. Resilience in this field requires continued spending on assets that appear duplicative, contingency exercises that may look excessive, and a political willingness to treat communications infrastructure as part of public safety rather than as a commercial afterthought. Many governments admire that logic in theory while underfunding it in practice.

The line went quiet in Tonga, and the larger lesson should not be softened by distance. When subsea cables fail, they reveal how much of state function has come to depend on physical systems that remain publicly underestimated precisely because they normally work so well. Climate stress does not need to be the sole cause of damage to make this strategically important. It is enough that changing marine and coastal conditions can complicate protection, narrow repair windows, and increase the cost of being wrong about redundancy. The unresolved question follows naturally. If communications are now as essential to emergency response and state continuity as roads or fuel depots, why do so many governments still treat the line beneath the water as a technical convenience rather than as one of the quiet frontiers on which authority either continues to function or falls silent?

Search and rescue as a strategic service

In August 2016, the cruise ship *Crystal Serenity* transited the Northwest Passage carrying more than a thousand passengers through waters long associated less with tourism than with remoteness, ice, and limited response reach. The vessel's operator had planned carefully, and Canadian authorities followed the voyage closely. The date is easy to verify. The place was the Canadian Arctic. What made the passage strategically revealing was not simply the ship itself, nor the excitement that accompanied its route. It was the question hanging behind it: if something went badly wrong, who could get there, how fast, with what aircraft, vessels, medical support, and command structure? The call has to connect.

That question captures why search and rescue belongs near the centre of any serious account of coastal and offshore governance under warming conditions. Search and rescue is commonly described in humanitarian terms, and rightly so. It exists to save life. But it is also one of the clearest ways a state demonstrates its practical presence in exposed regions. A flag on a map proves little when compared with an aircraft launched in difficult weather, a vessel diverted in time, a radio call answered by a system that knows what to do next, and a casualty brought to care despite distance, cold, or sea state. The call has to connect again, and not only technically. It has to connect institutionally.

This dual character is what makes search and rescue strategically important. It is both a life-saving service and a public test of competence. That test becomes sharper as activity increases in regions where the margin for rescue is narrow. More shipping, more tourism, more fishing, more offshore work, more local transport, and in some places more variable environmental conditions do not simply raise the number of potential incidents. They raise the burden on the systems that must respond when things go wrong. Crews are finite. Aircraft are finite. Fuel, maintenance windows, trained coordinators, medevac capacity, and communications coverage are finite. Search and rescue therefore sits at the junction between the state's humanitarian obligations and its administrative limits.

This is why search and rescue should not be treated as an afterthought to infrastructure, shipping, or offshore development. It is part of the operating condition that makes those activities governable. A state that permits or encourages activity in difficult waters is making an implicit promise about response, even when it avoids stating the promise too clearly. That promise need not guarantee success in every case. It does need to remain credible.

The concept to name here is demonstrable capability. Capability, in this setting, is not what exists in an inventory spreadsheet or a ministerial statement. It is what can be shown under pressure: crews mobilised, aircraft launched, communications sustained, casualties found, people recovered, information shared, and coordination held together across long distances and shifting conditions. The implication is simple and severe. Search and rescue does not merely support sovereignty. It performs it.

The Arctic and sub-Arctic make this especially visible because distance alters the meaning of response time. A rescue system in a densely served region may work with assumptions about nearby hospitals, multiple airfields, abundant vessels, redundant communications, and overlapping jurisdictions with considerable resource depth. In exposed northern waters, many of those assumptions weaken at once. Distances are longer. Weather may close quickly. Diversion options are fewer. Medical facilities are farther away. Spare aircraft may already be committed. Ports may be limited. Communications can be patchy. Search areas can be vast. Every one of these conditions can exist before any climate-linked change enters the scene. Warming matters because it can bring more activity into these environments and can alter ice, weather, and operational patterns in ways that increase demand on already thin systems.

The *Crystal Serenity* voyage drew attention precisely because it dramatised that burden. Even with careful planning, a large vessel in remote Arctic waters raises obvious questions about evacuation, medical care, reception capacity, and rescue reach. One successful voyage does not prove a region is fully ready for regular mass-tourism traffic any more than one severe incident proves it is ungovernable. What the example does show is that search and rescue capability is judged publicly through the plausibility of response. If the rescue question sounds uncomfortably open, then the political meaning of expanded access changes. A route may be navigable and still not be responsibly governable at the activity level the market imagines.

This is where increasing activity stretches systems in ways that are not always visible in official readiness language. Crews are among the first limits encountered. Search and rescue depends on skilled personnel whose training, experience, and judgement are difficult to replace quickly. Pilots, flight engineers, rescue swimmers, coordinators, maritime crews, medical teams, and technicians do not scale up overnight because vessel traffic or offshore interest has risen. If incidents remain infrequent and well spaced, a lean system may appear adequate. If two or three serious demands arrive close together, adequacy can vanish. Fatigue grows. Maintenance schedules tighten. Backup crews become primary crews. Small slippages in readiness become major constraints.

Aircraft availability compounds this problem. In many exposed regions, aircraft are the practical bridge between incident and response, yet they are especially sensitive to weather, maintenance, range, basing, and competing demands. A rescue helicopter cannot be in two places at once. A fixed-wing aircraft suitable for search may not be suitable for extraction. A medevac sortie can tie up the same network that maritime rescue depends on. Public debate often assumes that if aircraft are part of a national fleet, they are functionally present. This is rarely true. The relevant question is where they are based, what they can safely do in prevailing conditions, what range they can cover with margin, what crew availability looks like after previous sorties, and how quickly they can be released from other commitments. The call has to connect through those facts, not around them.

Coordination protocols are the quieter but equally decisive part of the system. Search and rescue succeeds or fails not only through assets, but through handoffs. Who takes the call? Which centre assumes lead responsibility? What information standard is used between maritime, aviation, military, health, and local authorities? Who can request outside support? Under what rules do neighbouring jurisdictions assist? How are local vessels, indigenous observers, coast guard units, and air assets brought into one picture quickly enough to matter? These questions may look procedural until the moment several agencies need to act under time pressure and uncertainty. Then procedure becomes capability. A well-designed protocol does not remove hazard. It reduces delay in the face of it.

A named example from Norway helps clarify the point. Search and rescue arrangements in the Norwegian High North are often regarded as comparatively serious because they have been shaped by long experience with harsh conditions, offshore activity, and the need for cooperation among different services. This does not mean the region is immune from stress or free from criticism. It does show that capability in exposed waters depends on more than equipment. It depends on repeatable coordination, pre-existing relationships, and systems that assume difficult rescue as a standing possibility rather than an unlikely exception. The lesson is not that every state can replicate Norway's approach exactly. It is that search and rescue credibility rests on the density and discipline of relationships around the assets, not on the assets alone.

That relational point matters because warming is often framed too simplistically in search and rescue discussions. Commentators sometimes imply that reduced ice or longer seasons make access easier and therefore rescue easier. Sometimes they do. They can also increase traffic, widen the area in which activity occurs, complicate forecasting, and expose systems to mixed conditions for which older assumptions are a poor guide. More vessels in waters that remain remote do not make those waters less remote. They make the consequences of remoteness more politically visible. The same applies to offshore platforms, research activity, and community transport. Activity can rise faster than response architecture matures. That is how strategic strain accumulates.

Search and rescue is also a test of legitimacy because it reveals whose safety is being counted seriously. A state that can mobilise impressively for a cruise vessel carrying affluent foreign passengers but struggles to provide equivalent urgency for local fishers, community transport, or Indigenous hunters exposes more than an operational gap. It exposes a hierarchy of attention. This is one reason local knowledge matters even in a chapter focused on offshore and coastal systems. People who travel these waters repeatedly often understand seasonal shifts, dangerous zones, communications gaps, and practical search constraints in ways formal systems need. Where they are integrated into reporting and response, capability improves. Where they are treated as peripheral, the state may possess aircraft and protocols while remaining blind in critical moments.

Still, a caution is needed. Search and rescue capacity should not be romanticised as though a perfect state can abolish distance. It cannot. Weather will still close. Sea state will still limit extraction. A vessel may still be too far away. A patient may still deteriorate before definitive care is reached. To speak of search and rescue as strategic is not to promise omnipotence. It is to insist that the seriousness of a state's presence can be judged by how honestly it plans for those limits, how well it communicates them, and how much effort it makes to narrow them where activity depends upon them. Overclaiming capability can be worse than admitting constraint, because visible failure after inflated assurance damages trust more deeply than acknowledged limitation does.

Concurrent crises reveal this harshly. A rescue system that performs well in single-incident drills may still be overwhelmed when several demands collide: a maritime distress call during wildfire evacuations, an offshore medical emergency during regional storm damage, an aircraft incident while communications are already degraded, a fuel delivery problem at the same time as a rescue sortie must be mounted from the same logistical network. These are not fanciful combinations. They are the sorts of compounding pressures climate-linked stress can make more plausible. Even well-designed systems can fail when concurrency strips away the spare margin they relied upon. This is why search and rescue belongs in a broader discussion of state capacity rather than in a silo of maritime heroics.

A fair counter-example deserves acknowledgment. Some search and rescue systems are far stronger than public pessimism suggests. Through regular exercises, dense interagency ties, local knowledge integration, and careful asset placement, they can achieve an impressive degree of reliability under difficult conditions. Success in hard cases is real and should not be dismissed. Yet even these systems reveal the central point. Their strength usually depends on sustained investment, training, maintenance, political attention, and a willingness to fund capacities that appear expensive relative to routine use. Their success is not evidence that search and rescue is a secondary or easily solved issue. It is evidence that credible capability is built, rehearsed, and paid for long before the distress call arrives.

Search and rescue also shapes diplomacy in quieter ways. States make claims about responsibility zones, offshore development, and maritime activity in language that implies an ability to govern those spaces. Other states, insurers, operators, and local populations judge those claims partly by whether rescue capacity appears believable. A successful rescue can reinforce confidence that a state is equal to the space it claims to manage. A visible failure can have the opposite effect, especially where distance and weather were already known constraints. Search and rescue thus becomes a form of signalling without the aggression associated with military posture. It is competence made visible under duress.

The humanitarian and strategic dimensions cannot be separated neatly. They share the same aircraft, crews, communication links, coordination centres, and public expectations. That is why search and rescue should not be relegated to the sentimental margins of security analysis. It is one of the clearest places where public authority is tested against geography. It is also one of the clearest places where a state can discover that its confidence in activity, routes, or offshore systems has run ahead of its capacity to answer a call from the edge. Capability demonstrations matter because they are remembered. Communities, mariners, operators, and foreign observers all watch the same event and draw conclusions about whether the system is real.

The call has to connect, and not only with the person in distress. It has to connect between aircraft and vessel, between local knowledge and central command, between weather information and launch decision, between rescue and medical handoff, between one incident and the next. That chain is where sovereignty becomes practical. The unresolved question is therefore not whether search and rescue is morally important; that is obvious. It is whether states expanding or tolerating more activity in exposed regions are investing in response systems proportionate to the promises that activity quietly makes, or whether they are relying on the hope that the next call will not arrive while the system is already busy proving itself somewhere else.

Chapter 6

Disaster Planning for Compounding Shocks

What fails first when the fire does not come alone? That question cuts closer to the governing problem than the more familiar language of preparedness usually allows. States often plan by hazard category because categories suit institutions: flood plans, wildfire plans, heat plans, blackout plans, evacuation plans. Each has a logic, a department, a budget line, a protocol. Real disruption is less obedient. Smoke arrives during heat. Power fails during evacuation. Roads close while medical demand rises. Shelters fill just as communications thin and rumours spread faster than official updates. The queue lengthens.

That queue is not only physical. It forms at fuel depots, emergency call centres, hospital triage desks, shelter entrances, procurement offices, and ministerial decision points. Compound shocks expose the same weakness in all of them: systems designed to absorb one strain at a time begin to compete with themselves. Staff are pulled in two directions. Vehicles serve one task and leave another uncovered. Backup plans assume services still exist elsewhere, only to discover that the neighbouring jurisdiction is under the same pressure. A disruption that appears sudden from the outside often reveals an older internal truth. Scarcity was already there. The overlap made it visible.

This is why compounding risk is best understood as a governance problem before it is treated as a forecasting problem. Hazards can overlap in time, place, or institutional load, but the decisive question is what those overlaps do to prioritisation. Who gets moved first when evacuation routes narrow and care needs diverge? Which service receives fuel when both the hospital and the water plant depend on backup generation? Which agency's information becomes decisive when several partial pictures conflict? Under single-event planning, such choices can be postponed behind the reassuring fiction that each system will function roughly as intended. Under compound strain, that fiction thins quickly. The queue lengthens again.

The central tension therefore lies between the administrative comfort of separate plans and the lived reality of interdependence. A government can possess competent plans for wildfire, power loss, smoke exposure, and emergency sheltering and still fail because those plans were not built to collide. Compounding shocks do not merely increase the volume of work. They alter the sequence through which failure travels. A shelter needs power. Power needs fuel. Fuel needs roads or deliveries. Roads depend on weather, traffic, and policing. Health services depend on all of them while also carrying their own staff shortages and care burdens. What looks like one emergency becomes a contest over shared dependencies. Resilience then stops being a virtue word and becomes a test of whether public systems can choose under pressure without pretending there is enough for everyone at once.

That test matters because public order in disaster does not rest chiefly on force. It rests on belief. People evacuate when they think the route still works, the shelter will hold, the hospital will receive them if needed, and the authorities are speaking from a real picture rather than issuing slogans into confusion. The same applies to institutional cooperation. Local officials share bad news early when they trust it will trigger support rather than blame. Utilities prioritise transparently when they believe the political system will defend necessary trade-offs. Households accept inconvenience more readily when they can see that scarcity is being managed rather than denied. Compound shocks are therefore social as well as logistical events. They reveal whether trust has been built deeply enough to survive shortage.

The stakes extend beyond the emergency itself. Systems that repeatedly face overlapping strain begin to change in quieter ways. Staff burn out. After-action reports grow more cautious. Temporary repairs become semi-permanent. Mutual aid agreements lose credibility if simultaneous regional demands keep exposing their limits. Public patience narrows when each crisis arrives with promises that sound familiar and lessons that seem not to have travelled into budgets or staffing. This is why compound disruption must be read not only as operational difficulty but as a measure of institutional learning. A state can respond visibly and still fail strategically if it cannot convert collision into reform before the next collision arrives.

The path ahead follows the logic of that collision. It begins with the blind spot created by single-event planning and moves into the harder terrain where evacuation, sheltering, and public order depend on whether systems under strain can still command cooperation. From there the focus turns to the service chains that make modern life look stable until one break degrades several others at once, before arriving at the politics of learning, where institutions reveal whether they can turn exposed weakness into changed practice rather than another polished summary of lessons observed.

What binds these themes is the simple but uncomfortable fact that compound emergencies punish administrative neatness. They do not respect the boundary between emergency management and public health, between infrastructure maintenance and social care, between communications strategy and logistics, between visible response and quiet prevention. A backup generator may work perfectly and still fail as a resilience measure if the fuel truck cannot get through. A shelter may be open and still not function if continuity of care collapses inside it. A route may remain passable and still cease to be a solution if misinformation drives too many people onto it at once. Systems fail through relation, not only through damage.

This is also why the chapter cannot treat planning as a matter of technical optimisation alone. Scarcity forces political choice, and political choice is judged afterwards through narratives of fairness, competence, and blame. Who was protected, who waited, who knew, who acted, who hesitated, who learned - these questions shape the next crisis before that crisis has even begun. The capacity to learn from compounding shocks therefore belongs alongside the capacity to respond to them. A state that performs heroically and learns badly may still be growing weaker. A state that responds less theatrically but changes its staffing, contracts, dependencies, and authority lines may be becoming stronger in the only sense that matters over time.

The queue remains the image worth keeping. Not because it is dramatic, but because it is ordinary and exact. A queue for beds, for fuel, for repair crews, for shelter, for information, for permission. Compound shocks begin to govern a society when too many of those queues begin to form at once and each one quietly depends on the others. The harder question, then, is not whether governments can imagine worse disasters. It is whether they can recognise which queue now reveals the real system limit before the rest begin to form behind it.

Compound events and planning blind spots

In June 2021, western Canada and the north-western United States endured an extreme heat event that strained hospitals, power systems, transport, and emergency services across a wide area. The place is easy to fix on a map. The date is easy to verify. The public often remembers heat as a single hazard, measured in temperature and death tolls. But the governing difficulty lay elsewhere: electricity demand rose as cooling became more urgent, transport systems behaved differently under stress, health services faced surging need, and in some places wildfire smoke either accompanied or soon followed the heat in ways that complicated every ordinary response. The queue formed quickly.

That queue is the right image because compounding shocks are best understood not as one dramatic event plus an unfortunate coincidence, but as overlapping claims on the same limited systems. A heatwave is one hazard. Smoke is another. A power outage is a third. Put them together and the problem changes category. People who are told to stay indoors during smoke may be trapped in homes that are dangerously hot if power has failed. Cooling centres cease to function if electricity or transport is disrupted. Hospitals treating heat illness must also manage respiratory distress from polluted air. Ambulances move through roads and communications systems already stretched. No single element needs to be unprecedented for the system as a whole to become unstable. The queue is what the overlap looks like in institutional form.

This is the essential meaning of a compound event. Hazards overlap in time, in place, or in the demands they place on public systems. Sometimes the overlap is obvious, as when flood and power loss arrive together. Sometimes it is administrative: a wildfire far from a city sends smoke into that city at the same time that heat drives demand for medical care and electricity upward. Sometimes it is social: one disruption empties staff reserves that another disruption had quietly assumed would still be available. The concept to name here is compounding load. The implication is exacting. Plans designed around single-event assumptions can fail not because their individual logic is foolish, but because they imagined the rest of the system remaining stable while one hazard was being managed.

This is why compound risk is fundamentally a governance issue. Physical overlap matters, but what turns overlap into crisis is prioritisation under scarcity. A government may possess a solid heat plan, a competent smoke advisory framework, and a plausible blackout contingency. What it often lacks is a worked-through answer to the harder question: when all three occur together, which populations, routes, facilities, and services receive the first margin of protection? Who gets power restored first when cooling is medically urgent and water pumping also depends on electricity? Which shelters remain usable when indoor air quality is compromised and transport is constrained? Which workers are redeployed, and what does their redeployment leave uncovered elsewhere? Compound events reveal that the real emergency plan is not the binder. It is the hierarchy of choices institutions make when several binders begin to contradict one another.

This contradiction usually arrives through ordinary dependencies rather than spectacular collapse. A backup generator appears to solve one problem until fuel delivery is disrupted. A shelter appears available until smoke control, staffing, and continuity of care reveal it is not suitable for many of the people told to use it. A hospital can expand emergency intake until air handling, staff fatigue, pharmaceutical delivery, and ambulance turnaround tighten at once. A utility may preserve the grid in aggregate while certain neighbourhoods, care homes, or water systems experience consequences far more severe than the official outage language suggests. The failure looks sudden because the public sees the moment of visible disruption. In reality the system was structurally primed.

A named example from California illustrates this logic with uncomfortable clarity. During recent years, severe wildfire smoke and dangerous heat have overlapped in ways that complicated public advice and strained multiple systems simultaneously. The problem was never simply that one hazard followed another. It was that the standard protective behaviours could collide. "Go to a cool place" becomes difficult advice if smoke makes outdoor movement risky, if cooling centres are limited, or if a power outage removes air conditioning where people are already sheltering. "Stay indoors" becomes dangerous if indoor temperatures become intolerable. Even where official plans exist for each condition separately, the overlap forces a different form of judgment. Public communication, staffing, shelter design, grid management, and health triage all have to be re-read together.

A second example can be seen in Texas during February 2021, where severe winter weather, electricity failure, water disruption, and cascading service breakdown exposed the brittleness of systems built on narrower assumptions than public confidence implied. The hazard set was different from smoke and heat, but the structural lesson is the same. Power was not merely one utility among others. Its failure degraded water treatment, heating, healthcare operations, communications, and household safety together. A disruption that might have remained severe but manageable became much harder to contain because critical services depended on one another more tightly than planning language had suggested. The event is relevant here not because every jurisdiction resembles Texas, but because it shows how quickly compound stress transforms infrastructure from sectors into a single field of vulnerability.

Planning blind spots often begin with categorisation. Modern administrations are good at dividing labour. Fire services manage one class of event, health services another, utilities another, civil protection another, local authorities another. This division is rational, even necessary. Yet it produces a chronic temptation to assume that when one system enters crisis mode, the surrounding systems remain broadly available as support. Compound events punish that assumption. A utility may need transport and policing support at the very moment transport and policing are overstretched by evacuation or traffic control. A hospital may need generators just as the fuel distribution network has narrowed. Emergency shelters may need medical staffing just as clinics are short-handed and roads are congested. The blind spot lies not in ignorance of hazards, but in confidence that the support environment will remain intact long enough for sectoral plans to work.

This is why staffing is so often the first hidden point of failure. Public attention gravitates towards damaged infrastructure and visible shortages, yet compound events commonly begin to defeat institutions through people. A workforce sized for one emergency can become dangerously thin when the same staff are needed in more than one place, when smoke or heat affects the workers themselves, when schools close and caring responsibilities rise, or when the emotional and physical burden of extended response begins to slow judgment. A plan that assumes "additional staff will be redeployed" may be administratively tidy and operationally false. The staff may already be committed, unavailable, or too exhausted to function at the level the plan imagines. Resilience often fails on the rota before it fails on the map.

Supply chains behave similarly. A single hazard may be survivable with existing reserves, routine deliveries, and familiar improvisation. Multiple overlapping hazards place those arrangements under a different type of pressure. Fuel is needed for generators, evacuation vehicles, hospitals, telecoms sites, and emergency operations centres at the same time. If roads are degraded, smoke reduces visibility, or power outages affect depots and payment systems, allocation becomes a political act. Supply chains are then exposed not as neutral pipelines but as systems that embody prior decisions about what is scarce, what is buffered, and who will wait. A government that has never forced itself to map these dependency chains in advance is likely to discover them in public, under blame.

That blame matters because compound events are fertile ground for distrust. People comply with difficult instructions when they believe the system still holds. They leave early when they think routes will remain viable and destinations usable. They tolerate rationing when they think allocation is intelligible and temporary. They share space, queue, and inconvenience when they believe others are subject to the same rules and that the authorities know what they are doing. Compound events unsettle these beliefs because every visible contradiction in official guidance invites suspicion. If one agency says stay indoors, another says seek cooling, and a third announces outages with no clear priority plan, citizens do not merely feel confused. They begin to revise their judgment of the state.

Misinformation becomes more potent in precisely such conditions, not because publics are inherently gullible, but because ambiguity widens the space in which unofficial guidance can outperform official vagueness. During overlapping smoke, heat, and power stress, rumours about shelter availability, evacuation routes, fuel shortages, or service collapse can move faster than corrected information if institutions are slow, fragmented, or overconfident. This is one reason compound risk cannot be treated as a purely technical planning matter. It reshapes the information environment. People do not wait patiently for integrated government to appear. They act on what seems timely, plausible, and socially confirmed. If the official system cannot supply that, compound hazards begin to colonise public reasoning as well as public services.

Still, a caution is necessary. Not every overlap produces systemic failure, and not every multi-hazard episode deserves grand language about compounding shocks. Some jurisdictions absorb several simultaneous pressures reasonably well because they have spare capacity, strong interagency relationships, decentralised decision rights, and a habit of planning around dependencies rather than sectors alone. Some overlaps remain operationally significant but politically containable. It would be a mistake to inflate every difficult week into proof that standard emergency planning is obsolete. The stronger argument is narrower. Where hazards overlap enough to compete for the same staff, shelter, logistics, or public compliance, single-event planning becomes less reliable and governance quality matters more.

That limitation is important because it points to the right kind of seriousness. The goal is not to replace all hazard-specific planning with amorphous multi-threat rhetoric. Hazard-specific planning remains indispensable. Fire behaves differently from flood; heat differs from outage. The point is that institutions need to know where those plans will collide. Which assumptions about staffing are shared across plans? Which routes are claimed by more than one emergency function? Which facilities are supposed to serve as both shelter and treatment space? Which backup systems depend on the same fuel, communications, or maintenance capacity? These are not theoretical questions. They are the points at which a compound event becomes visible to governance.

A quiet motif runs through all of this: the queue. A queue for ambulances outside an emergency department treating heat stress and respiratory illness at once. A queue for fuel when generators, evacuation vehicles, and water systems depend on the same shrinking deliveries. A queue for shelter beds that were designed for one disruption and meet another at the door. A queue for authorisation as agencies discover their plans overlap but their authority does not. The queue is useful because it translates compound risk from abstraction into a visible sign that several dependencies have begun to press on the same bottleneck. When the queue lengthens, one is already looking at governance, not just hazard.

This is why compounding shocks should be read as tests of public choice under pressure. They force institutions to declare, through action, which risks were treated as manageable in theory because someone assumed the second hazard would wait its turn. Climate-linked disruption matters here not only because it may intensify individual hazards, but because it increases the plausibility of overlapping strain across systems already optimised too tightly to absorb much surprise. The strategic question is therefore more demanding than "are we prepared for fire, heat, smoke, outage, or flood?" It is whether the system knows which queue to shorten first when several begin to form together and each one quietly decides the length of the others.

Evacuation, shelter, and public order

In August 2023, authorities in the Northwest Territories ordered the evacuation of Yellowknife as wildfire threatened the city. The place was northern Canada. The date is easy to verify. The order itself was only the beginning. The real test started the moment people began to move: roads had to hold, aircraft had to be available, fuel had to reach vehicles, shelters had to receive not just bodies but needs, and public messages had to remain credible as conditions changed. The route had to hold.

That phrase captures the deeper problem of evacuation under compounding stress. Evacuation is often imagined as movement from danger to safety, as though danger occupies one place and safety another. Real evacuations are messier. Safety is assembled while people are already in transit. Roads remain open until they do not. Transport plans assume vehicles, fuel, weather, traffic control, and communications will align. Shelters look adequate until continuity of care, medication, disability access, food supply, indoor air quality, or staffing reveals a narrower truth. Public order, meanwhile, is not maintained by command alone. It depends on whether people believe the system still functions. The route had to hold again, and so did the confidence around it.

This is why evacuation, shelter, and public order belong in the same analytic frame. They are not separate emergency topics that happen to occur in sequence. They are one chain of legitimacy. An evacuation order without a credible route is not only operationally weak; it is politically corrosive. A shelter network without continuity of care is not simply under-resourced; it can teach the public that official safety categories do not match real vulnerability. Public order without trustworthy information is not order for long. The concept to name here is compliance infrastructure. The implication is plain: people comply when the state demonstrates that movement, reception, and care are more than instructions shouted into uncertainty.

The practical limits of evacuation begin with routes, but they do not end there. A road that appears open on a map may already be compromised by congestion, smoke, flood, fallen debris, or fuel scarcity. A flight plan may exist while aircraft availability remains thin, weather unpredictable, and priority rules contested. A town may be told to leave while neighbouring jurisdictions are also under strain, reducing the receiving capacity that evacuation quietly assumes. This is where compound shocks expose single-event planning. A route designed for one hazard can become unworkable when another hazard narrows visibility, overloads fuel systems, or fills hospitals and shelters before evacuees arrive. The public sees a traffic jam or a delayed flight. The state is confronting a layered system limit.

The 2016 Fort McMurray wildfire evacuation in Alberta remains instructive because it showed both the scale of movement possible under acute pressure and the fragility of the systems supporting it. The event is often remembered for fire intensity and mass departure. It should also be remembered for what it revealed about route dependence, fuel logistics, communication, and trust in public instruction. An evacuation of that magnitude is never just about road width. It depends on whether authorities can sequence messages, maintain enough order that people do not all make the same desperate choice at once, preserve access for emergency vehicles, and support those who cannot self-evacuate easily. The visible motion of traffic conceals the administrative labour beneath it.

Continuity of care is where the moral seriousness of sheltering becomes operational. Emergency shelters are too often imagined as neutral spaces containing food, beds, and basic safety. For many evacuees they are also sites where chronic illness, disability, pregnancy, mental distress, substance dependence, respiratory vulnerability, and the needs of children or older people intensify rather than pause. A shelter that cannot handle refrigeration for medicines, oxygen support, accessible sanitation, private family needs, or trauma-sensitive staffing may still count as open while failing a significant share of those told to rely on it. This is not a marginal problem. It is one of the ways public trust is either built or broken in real time.

Hurricane Katrina in 2005 remains one of the clearest historical warnings about what happens when evacuation, shelter, and public order are treated as separable functions. New Orleans is the place. The date is easy to verify. Much of the analysis of Katrina has focused, rightly, on levee failure, emergency management breakdown, and the scale of human suffering. Yet for the purposes of this chapter, the enduring lesson lies in the collapse of confidence linking official instruction to actual protection. Routes proved inadequate or inaccessible for many, shelter conditions became sites of exposure as well as refuge, and the public meaning of order changed rapidly when institutions appeared unable to deliver the care and coherence their directives presumed. The lesson is not that every evacuation resembles Katrina. It is that public order degrades quickly when people conclude that movement does not lead to dependable safety.

That degradation is often described too loosely as panic. Panic is an easy word, and often a lazy one. Most people in disaster do not behave irrationally. They behave under uncertainty, incomplete information, and unequal resources. They test official guidance against what they see, what neighbours say, what social media reports, and what prior experience has taught them about whether systems really hold. If authorities say shelters are ready but previous shelters lacked medication support, many will hesitate. If officials promise that roads are clear while people receive messages or images suggesting otherwise, departure patterns become less cooperative. If fuel is scarce or wrongly rumoured to be scarce, self-protective behaviour can overwhelm careful planning. Public order is therefore best understood not as obedience secured by force, but as coordination sustained by credible expectation.

Misinformation exploits precisely that expectation. It does not need to invent an entirely false world. It only needs to widen doubt at the moment when trust is already under strain. During evacuations and shelter operations, misinformation about road closures, shelter conditions, criminal threat, fuel availability, hospital capacity, or preferential treatment can accelerate breakdowns in cooperation even when the underlying system is still functioning tolerably well. This is why information management belongs inside disaster operations rather than at the edge of them. A false rumour that one shelter is full may divert hundreds of people to a less suitable site. A misleading claim that routes are closing can trigger simultaneous movement that clogs the only viable exit. The route has to hold materially and informationally.

This is also why resilience must be treated as a social and administrative achievement rather than a physical property. A community is not resilient because it has a designated shelter and an evacuation map. It is resilient when people trust those designations enough to act before compulsion becomes chaos, when local officials and residents understand how to prioritise care for those with the least mobility and greatest dependency, and when state agencies can revise guidance without destroying confidence in the guidance already issued. Resilience, in this setting, is visible behaviour: buses arriving for those without cars, pharmacies or care records integrated into shelter operations, fuel prioritised for evacuation and emergency services, local radios or trusted leaders echoing official information without contradiction, and police presence used to facilitate movement rather than merely to project control.

A counter-example is worth keeping in view because not all mass movement under climate-linked stress descends into breakdown. Some jurisdictions have shown that orderly evacuation and sheltering are possible when pre-existing relationships, transport planning, public communication, and care integration are taken seriously. Parts of Australia during bushfire response, and some hurricane evacuations in the United States, have demonstrated that large-scale movement can proceed with substantial public cooperation when messaging is trusted and systems are visibly prepared for differentiated needs. This matters because it prevents the argument from turning melodramatic. The point is not that evacuation necessarily fails under compound stress. It is that success depends on social and administrative conditions, often described too briefly or funded too thinly to survive repetition.

Still, even strong systems can be overwhelmed by concurrency. A well-designed shelter plan may assume additional staff from neighbouring areas, only to find that those areas are handling smoke, heat, or power disruption of their own. An evacuation route may work under one threat and fail when a second hazard narrows visibility or disrupts fuel supply. Hospitals prepared for incoming evacuees may lose effective capacity if their own generators run low or their staff cannot commute. Compound shocks expose the false comfort of isolated planning assumptions. The issue is not whether one plan was badly written. It is whether all the plans together assumed a degree of system stability that no longer exists under layered pressure.

This is where continuity of care deserves further emphasis. Too many emergency systems still treat evacuees as generic units to be counted, moved, and housed. Real people arrive with oxygen needs, dialysis schedules, prescription gaps, children needing supervision, older relatives who cannot navigate crowded dormitory conditions, animals on which livelihoods depend, language barriers, and trauma histories that can make shelter conditions unbearable. If these realities are not integrated into planning, the state effectively tells vulnerable people that compliance requires entering a system not built for them. Many will then delay leaving, seek informal alternatives, or disappear from the official picture. What gets labelled non-compliance may in fact be rational refusal of an unsuitable option.

Public order is deeply affected by that mismatch. Order does not crumble only when crowds become unruly. It also degrades when many small acts of justified improvisation accumulate because official pathways are too blunt to carry the real variety of needs. Families bypass shelters because they fear losing continuity of care. Drivers leave early or late because official timing does not match local reading of the road. Community networks circulate their own advice because state information is too generic or slow. Informal systems can save lives. They can also complicate prioritisation when they develop in opposition to public systems rather than in tandem with them. The best disaster planning therefore does not aim to eliminate informal adaptation. It aims to make official systems credible enough that adaptation works with them instead of around them.

The role of local trust networks is especially important here. Community leaders, local health staff, school officials, faith groups, Indigenous authorities, and volunteer organisations often function as translators between official plans and lived reality. They know who lacks transport, who depends on electricity for medical reasons, who will not use a mixed-gender shelter, who needs medication support, who mistrusts government warnings for good historical reasons, and which rumours are gaining ground. Treating these actors as peripheral can turn a technically strong evacuation order into a socially weak one. Treating them as part of the operational architecture can preserve compliance where central messaging alone would fray. This is not sentimentality. It is governance using the channels through which real populations move.

The politics of visibility complicates all of this. Governments are often rewarded for highly visible response and less rewarded for the slow, expensive, and easily overlooked work of making shelters medically and socially usable, keeping evacuation routes fuelled, or rehearsing contradictory hazards in one exercise. As a result, the institutional temptation is to emphasise command, movement, and headline-capable action. Yet the strategic burden sits in the quieter details. Did the shelter accept people with complex care needs? Did the buses arrive? Did the backup generators have secure fuel? Were police, health staff, and local volunteers operating from the same information? Did people return home with enough continuity that the evacuation did not become a second crisis? These are the questions by which competence is remembered.

A further limitation should be admitted. No government can eliminate all hardship or guarantee perfect sheltering under severe compounding stress. Resources are finite. Conditions change. Some evacuation orders will arrive late because the hazard moves faster than forecast confidence. Some shelters will operate under painful compromise. Some misinformation will spread no matter how disciplined official communication becomes. Recognising these limits is not a retreat from responsibility. It is a way of defining it more honestly. The relevant measure is not whether hardship occurs, but whether institutions have reduced avoidable hardship by designing systems around real dependencies rather than administrative convenience.

That is why evacuation, shelter, and public order deserve to be understood as one strategic service rather than three adjacent tasks. They reveal whether a state can convert warning into movement, movement into safety, and safety into continued social cooperation while conditions are still changing. If any link fails, the others weaken quickly. The route closes. The shelter loses legitimacy. Rumour gains force. Police and emergency staff are forced to compensate for losses in trust they cannot restore by presence alone. The costs of earlier administrative shortcuts then arrive all at once.

The route had to hold in Yellowknife, as it has had to hold in many places before and since. But the deeper question is not only whether the road remains passable or the shelter doors remain open. It is whether the public, when told to move under pressure, sees a state that has prepared for the full human reality of movement, care, and return - or a state that still imagines evacuation as the easy part because it has confused issuing an order with being able to keep the promise hidden inside it.

Critical services under stress

In February 2021, when severe winter weather struck Texas, the public quickly understood that the electricity system was in trouble. The place was Texas. The date is easy to verify. Power outages spread across a large area, but the deeper failure was not electrical alone. Water systems faltered, heating became unsafe or impossible in many homes, hospitals faced difficult operating conditions, fuel distribution was disrupted, and emergency response moved through a landscape in which several supposedly separate services had started to fail together. One switch had tripped. Many others followed.

That sequence matters because critical services are often governed as if they were adjacent sectors rather than interdependent systems. Electricity belongs to one field of expertise, water to another, health care to another, fuel logistics to yet another. Each has its own regulators, operators, maintenance language, emergency plans, and budget lines. Under ordinary conditions that division feels manageable, even efficient. Under compound stress it becomes misleading. A hospital is not merely a health facility; it is an electricity user, a water user, a fuel user, a communications node, a staffing system, and often a shelter of last resort. A water treatment plant is not merely a utility; it depends on power, chemicals, transport, telemetry, maintenance, and public compliance. The switch had tripped again, this time in administrative thinking.

This is why loss of one critical service so often degrades several others in quick succession. The deterioration looks sudden because the public experiences the moment when visible continuity breaks. In reality, the system was already structurally primed for spiral. A backup generator had fuel for a time but no guaranteed resupply. A fuel depot relied on electricity for pumping or payment systems. A telecoms site had battery support but not for as long as the outage lasted. A hospital could continue elective restraint for a while until staffing, air handling, sterile supply, and patient transfer routes tightened at once. The concept to name here is dependency cascade. It matters because it shifts analysis away from isolated outage toward the chain through which one outage becomes several institutional emergencies.

The chain is often shorter than officials prefer to admit. Electricity fails, and water pressure begins to degrade because pumping and treatment depend on power. Water systems weaken, and hospitals face sanitation, sterilisation, and cooling problems. Communications degrade as backup power at key sites runs down or fuel deliveries slip. Fuel distribution, assumed to be the answer to backup power problems, proves dependent on road access, functioning depots, payment systems, and tanker availability. Public communication then becomes harder at exactly the moment when officials need people to conserve, move, or remain calm. What appears as a rapid collapse is frequently a latent design choice exposed by stress. Efficiency has trimmed the margin. The switch had tripped long before anyone saw the darkness.

The Texas crisis is instructive precisely because it makes interdependence hard to deny. Power failure was central, but power failure alone does not explain the wider social and institutional disruption. Water treatment and delivery systems encountered serious difficulty. Heating failures magnified health risk. Fuel and mobility constraints complicated response. Hospitals and emergency services operated under increasingly severe limitations. The strategic lesson does not depend on the precise details of every technical breakdown. It lies in the fact that no critical service existed alone. Each service had been planned and governed with enough sectoral confidence to obscure the speed with which one failure would undermine the rest. A state can possess plans for each system separately and still be strategically surprised by how fast they lean on one another once stress begins.

A second example comes from Hurricane Maria in Puerto Rico in September 2017. Public attention focused, understandably, on the storm and the prolonged electricity outage that followed. Yet the deeper pattern was a multi-service degradation in which power loss affected water systems, healthcare operations, communications, refrigeration, transport, and fuel use across the territory. A hospital running on backup generation was not secure simply because its generators functioned. It needed fuel deliveries, reliable maintenance, staff transport, sterile supplies, and functioning referral pathways. Water systems did not merely need pipes and pumps; they needed electricity, operators, treatment inputs, and communications.

The storm was the initiating hazard. The system-wide emergency emerged through dependency chains. One switch had become many.

This is where the argument becomes a method rather than a warning. To plan for critical services under stress, institutions need a way to map dependency chains before crisis reveals them publicly. The method begins with the service apparently at the centre of concern: a hospital, water plant, telecoms node, fuel depot, emergency shelter, or transport hub. The next question is not whether that service has a plan. It is what the service absolutely requires to function for the next hour, the next day, and the next week. Electricity may be the first answer, but rarely the only one. Water, fuel, staff presence, data links, pharmaceuticals, spare parts, security, waste removal, transport access, and legal authority often follow. Then the method asks which of those inputs are single points of failure, which depend on the same supplier or route, which have meaningful redundancy, and which degrade invisibly before they fail visibly. Planning becomes a matter of tracing chains, not merely checking assets.

This sounds obvious when written out. It is often neglected in practice because institutions tend to map dependencies only within their own operational horizon. A hospital may know its generator capacity but not the fragility of the regional fuel delivery network. A fuel planner may know depot inventories but not the electricity and communications dependencies built into distribution. A water authority may know treatment vulnerabilities but not the hospital consequences of a boil-water notice issued during a heat emergency. Each actor sees its own diagram. Very few see the combined picture. The result is a class of emergencies that appear to accelerate magically when in fact the acceleration has been built into the arrangement.

Backup generators offer the clearest example because they are so often invoked as proof of resilience. A building, facility, or tower has backup generation, therefore continuity is secured. That statement can be true for minutes, hours, or days. Its value depends on everything that follows. How much fuel is on site? How quickly can it be replenished? Can the delivery vehicle reach the site if roads are flooded, iced, blocked, or under curfew? Does the fuel supplier itself have power and staffing? Are transfer pumps functioning? Does the generator support the whole facility or only selected systems? Has it been tested under full operational load? Are spare parts and technicians available if it fails after prolonged use? The phrase "backup power" is often treated as the end of the analysis. It should be the start of a harder one.

This is where the example of fuel-constrained backup generation becomes especially revealing. Imagine a hospital, water plant, or communications centre that loses grid power during a heat emergency or smoke event. The backup generator starts and the institution is briefly reassured. Then a different dependency enters: fuel delivery. If tanker access is hindered by road closure, labour shortage, competing emergency demand, or disrupted depot operation, the generator becomes a timed promise rather than a secure capability. Officials may continue to describe the service as operational while the clock is already running down. Public messaging may not catch up until the margin is almost gone. The problem, again, is not that the generator was useless. It is that the dependency chain beneath it was treated as someone else's field.

Compound events make these chains harsher because they add simultaneous demand to already narrow margins. During wildfire smoke and heat, for example, electricity demand may rise because cooling becomes more important even as public advice encourages indoor sheltering. If the power system falters, backup generation becomes more critical across hospitals, care homes, telecoms, and water services at exactly the same time. Fuel distribution then faces multiple priority claims while road safety, air quality, and labour conditions may all be worsening. Under these conditions, the failure of one service does not merely pressure others sequentially. It alters them all at once. What looked like redundancy becomes competition over a shared lifeline.

This is why critical service planning must be as social as it is technical. Health services rely on staff getting to work, patients being moved appropriately, carers being reachable, and public instructions being followed. Water systems depend on public compliance when restrictions are imposed. Fuel distribution depends on orderly demand, not panic purchasing or rumour-driven diversion. Electricity restoration depends partly on whether households, businesses, and institutions understand prioritisation and respond accordingly. Trust is therefore not a civic afterthought. It is one of the things that keeps dependency chains from breaking faster than the hardware alone would dictate. A public that believes the system still functions will often conserve, queue, or comply in ways that preserve capacity. A public that stops believing this can unintentionally accelerate the collapse it fears.

The same is true inside institutions. Interdependency becomes much harder to manage when organisations mistrust one another's data, timelines, or priorities. A hospital may receive optimistic restoration estimates from a utility and plan accordingly, only to find those estimates revised too late for safe adjustment. A municipal authority may assume fuel deliveries are protected, while transport officials are already reallocating routes under emergency conditions. A water utility may not communicate quickly enough with health services about degraded pressure or treatment limits. None of these failures requires malice or incompetence in the crude sense. They arise because interdependent systems often operate through separate hierarchies until crisis forces them into one another's field. The spiral looks sudden because cooperation has not been practised under the same strain.

A fair limitation must be acknowledged. Not every critical service failure produces a broad cascade, and not every interdependency is equally consequential. Some systems are better buffered than they appear. Some facilities have true redundancy rather than cosmetic backup. Some outages remain local or short enough that downstream effects do not intensify. It would be a mistake to treat every disruption as the beginning of inevitable spiral. That would simply replace one form of complacency with another kind of exaggeration. The better lesson is discriminating rather than apocalyptic: which services are genuinely foundational, which dependencies are tightly coupled, which buffers are real, and which assumptions about restoration speed or substitute provision have never been tested against compound stress.

A useful counter-example lies in institutions that have deliberately mapped and exercised these chains. Some hospitals, utilities, and emergency management systems do not merely maintain backup power and contingency documents; they run joint exercises around fuel supply, transport constraints, communications degradation, and staffing shortages. They treat continuity not as an internal asset problem but as a network problem. This matters because it shows the argument is not fatalistic. Dependency chains can be understood better and, in some cases, loosened through redundancy, revised routing, better data sharing, decentralised stockpiles, and more honest public communication. The point is not that critical services are doomed to spiral together. It is that they already can, and often do, when planning remains trapped inside sector boundaries.

Mapping dependency chains, then, is a practical discipline of attention. Start with the service one wishes to protect. Ask what it needs to function immediately, then what those needs depend upon in turn. Identify where two or more supposedly separate services rely on the same route, depot, data system, contractor, or staffing pool. Distinguish between hard dependencies, where loss halts function, and degradational dependencies, where loss reduces quality, speed, or safety without immediate collapse. Ask which failure would be publicly visible first and which would be institutionally decisive first; they are often not the same. Then ask which assumptions about recovery time, replacement, or mutual aid sit beneath the whole picture. This is how a state begins to see spirals before the public has to live inside them.

The political significance of this method should not be understated. Governments often prefer neat infrastructure stories because neat stories fit ministerial portfolios and budget codes. Interdependence complicates accountability. If hospitals fail because fuel deliveries falter during a power outage and water systems are simultaneously degraded, who exactly owns the problem? In a narrow administrative sense, several actors do. In a strategic sense, the state as a whole does, because the public does not experience service failure as a seminar in jurisdictional complexity. It experiences a system that either held or did not. That is why interdependence is a governance problem before it becomes an explanatory one. The chain may be technical. The judgement is political.

The switch had tripped in Texas, and long before that in other places where services once treated as separate turned out to be one another's conditions of possibility. The hardest question is not whether critical systems are connected; that is now beyond doubt. It is whether governments are willing to plan, budget, and communicate as though a hospital, a water plant, a telecoms node, and a fuel depot are part of the same emergency organism - or whether they will go on waiting for the next "sudden" spiral to reveal what the diagrams should already have shown.

Learning systems and after-action politics

In August 2005, Hurricane Katrina struck the Gulf Coast of the United States and exposed failures so visible that entire shelves of reports, commissions, inquiries, and recommendations followed. The place was New Orleans and far beyond it. The date is fixed. The event remains one of the most studied disasters in modern administrative history. Yet one of its most durable lessons is not merely that institutions failed, but that institutions are often remarkably skilled at describing failure afterward without altering the incentives that produced it. The file is written.

That is the right place to begin because post-disaster learning is too often imagined as a natural consequence of suffering. A crisis occurs, evidence is gathered, lessons are identified, reforms follow. The sequence sounds rational and morally satisfying. In public administration it is frequently false. Lessons are not scarce. Implementation is. Most governments can produce after-action reviews, expert reports, parliamentary hearings, audit findings, or internal debriefs after a major emergency. What they struggle to produce is the slower, less visible labour of changing staffing models, procurement rules, maintenance schedules, interoperable communications, reserve practices, training routines, and decision rights before the next disruption arrives. The file is written again, often more elegantly than before.

This gap between identification and implementation is not a bureaucratic accident. It is structured by incentives. Political systems reward visible response more readily than quiet prevention. A minister standing at a podium after a flood, fire, blackout, or evacuation can show urgency, empathy, and command. The budget line that would have reduced the severity of the crisis months earlier often looked optional, expensive, or vulnerable to cuts because its success would have appeared as non-event. Public administration inherits this asymmetry. Staff can be commended for working through the night in response. They are rarely celebrated for maintaining a reserve contract, testing a backup fuel route, revising a shelter protocol, or insisting on a more cumbersome but less brittle system design. The crisis has a stage. Prevention mostly has a ledger.

This is why institutional learning should be treated as a strategic capability rather than an ethical aspiration. Capability can be observed. It can be measured in whether after-action findings alter the next budget, whether exercises produce revised staffing patterns, whether mutual aid assumptions are tested against simultaneous stress, whether lessons about generator fuel dependency change depot planning, whether evacuation reports lead to different shelter design, whether data-sharing failures lead to new authority or simply a new paragraph in the executive summary. The concept to name here is follow-through capacity. It matters because without it the state becomes eloquent about its weaknesses and structurally committed to repeating them.

Show the pattern in a familiar disaster and it becomes difficult to deny. After major emergencies, reports often identify remarkably similar categories of problem: fragmented communication, poor interoperability, inadequate staffing depth, unclear authority, fragile logistics, vulnerable populations poorly integrated into planning, procurement delay, unrealistic assumptions about mutual aid, and failures to translate monitoring into early action. None of this is mysterious. Much of it is already known somewhere in the system before the event. The question is why those known weaknesses so often survive into the next crisis. The answer lies partly in politics, partly in administration, and partly in the relationship between the two. Learning competes with normal business, and normal business is usually organised to postpone the awkward changes learning requires.

A named example from the United Kingdom illustrates the point in a domain adjacent to climate disaster but highly relevant to emergency governance. The public inquiry into the Grenfell Tower fire, which followed the June 2017 disaster in London, revealed not simply a chain of immediate failures but a long accumulation of warnings, assumptions, and institutional habits that had not been translated into effective preventative change. Grenfell is not a climate event, and it should not be folded carelessly into a climate narrative. Its relevance here lies in the politics of after-action knowledge. It shows how systems can contain warnings, guidance, fragmented accountability, and visible concern without producing the kind of integrated follow-through that would alter risk materially. The lesson is not identical across domains. It is soberingly familiar.

Disaster planning under compounding shocks suffers from this problem in a particularly acute form because many of its most important lessons are cross-sectoral. A wildfire review may reveal shelter weaknesses that belong partly to health care, partly to transport, partly to social services, and partly to public communication. A blackout inquiry may uncover fuel logistics problems that touch utilities, road management, police escort priorities, and private suppliers. A flood report may expose hospital continuity failures that depend on water, data, staffing, procurement, and backup power. The more a lesson crosses institutional boundaries, the less likely any one organisation is to own it fully. Responsibility diffuses. The file is written by many hands and implemented by none.

Blame shapes this process at every stage. After-action reporting takes place inside a political environment where institutions want to appear candid but not reckless, self-critical but not legally vulnerable, transparent but not professionally suicidal. These are not trivial pressures. Officials know that plain acknowledgement of preventable weakness can trigger litigation, reputational harm, budgetary punishment, or leadership change. The result is a style of reporting that often names deficiencies in terms broad enough to sound serious and vague enough to spread accountability thinly. "Communication gaps", "lessons identified", "need for improved coordination", and "opportunities to strengthen resilience" all have their place. They can also become the grammar of defensive learning. The report exists. The decision architecture remains remarkably similar.

This is where visible response becomes politically dominant. Governments and agencies often find it easier to promise better crisis command than to fund the mundane systems that make command less necessary. More training exercises, more public statements, more review panels, more emergency powers, more dashboards - each offers the appearance of learning. Some of these investments are real improvements. Many leave underlying fragilities largely intact. The hard changes are usually quieter: more warehouse space, more staff in overlooked technical roles, slower but more realistic procurement, buffer stocks that appear inefficient in ordinary years, revised building maintenance, updated sheltering standards, more interoperable communications, clearer delegated authority, agreements that assume neighbouring jurisdictions may also fail. These are harder to celebrate because they lack drama and often offend managerial ideals of efficiency.

The politics of budget cycles worsens the problem. After a disaster, funds may briefly become available. Windows for reform open. Public attention is intense. Yet the institutional machinery required to translate urgency into durable change is slow. Procurement rules still apply. Hiring takes time. Competing priorities return quickly. Political leadership changes. Media attention moves on. The departments responsible for implementation may not be the departments that felt the reputational heat most sharply. Within a year or two, the event can already be drifting from immediate administrative memory into a curated archive of lessons. A strategic learning system resists that drift. Most institutions do not.

A second named example shows the problem more squarely inside disaster management. After Hurricane Maria struck Puerto Rico in September 2017, a wide range of reports, investigations, and reviews documented failures and vulnerabilities across electricity, health care, communication, logistics, and emergency management. Much of what they described was highly specific and operationally relevant. The enduring question, however, has not been whether lessons were articulated. It has been whether those lessons were translated into funding decisions, infrastructure redesign, preparedness practice, and governance reform durable enough to matter under future compound strain. This is the core problem of after-action politics. Reporting can be intense, sincere, and widely publicised while institutional change remains partial, delayed, or vulnerable to later retrenchment.

This is why a strategic approach to learning demands different indicators from the ones public discourse usually prefers. The useful question is not whether a jurisdiction held a review, published findings, or announced a resilience agenda. The useful question is what happened next in forms that can be checked. Did emergency shelter guidance change and get funded? Were backup systems reconfigured after fuel or power dependencies were exposed? Were staffing shortages addressed in the actual roster, not only acknowledged in a report? Did mutual aid agreements get rewritten to account for simultaneous regional strain? Were after-action recommendations assigned to specific authorities with deadlines, budget backing, and public tracking? Did the next exercise simulate the failure revealed by the previous real event, or revert to a tidier scenario? Learning that cannot be traced into decisions is not meaningless, but it is strategically thin.

The role of institutional culture should not be ignored. Some organisations treat candour as a sign of seriousness. Others treat it as an invitation to punishment. In the first, near-misses and uncomfortable findings are more likely to surface early enough to prompt adjustment. In the second, reporting becomes cautious, selective, or ceremonial. Disaster systems that depend on blame to enforce accountability often discover that blame also suppresses the raw material of learning. This does not mean abandoning accountability. It means separating two tasks that public debate often muddles together: identifying who must answer for failure, and building conditions under which future warnings and internal criticism are not silently discouraged. These tasks can coexist. Many institutions have not learned how to make them coexist well.

A useful counter-example lies in domains such as aviation safety, where reporting cultures, standardised investigation, and procedural follow-through have in some contexts produced stronger traditions of learning than many disaster management systems enjoy. The comparison is not perfect. Aviation incidents are often more bounded, more technically standardised, and more tightly regulated than climate-linked compound emergencies. Still, the contrast is revealing. Where investigation is expected to produce operational change, where recommendations are tracked, and where near-misses matter institutionally rather than rhetorically, learning behaves less like moral commentary and more like system maintenance. The point is not that disaster governance can simply copy aviation. It is that serious learning requires structures that extend beyond the publication of findings.

This also explains why learning must be assessed over time rather than in the emotional wake of an event. Immediately after crisis, institutions often appear highly reflective. They commission reviews, thank responders, promise reform, and acknowledge hard truths. The harder evaluation comes later. Did later budgets preserve or cut the relevant capacities? Did leadership turnover stall implementation? Did agencies defend preventive spending when the memory of crisis had cooled? Were politically awkward recommendations quietly buried because they implied service withdrawal, revised standards, higher costs, or uncomfortable admission of prior neglect? Strategic learning can survive the news cycle. Performative learning usually cannot.

The phrase "lessons learned" deserves suspicion for this reason. It frequently means only that lessons have been identified, discussed, or filed. Learning, in any serious sense, means altered behaviour under comparable conditions. A system has learned from a fuel-constrained generator failure when the next continuity plan includes real fuel logistics, not just a stronger sentence about backup power. It has learned from shelter breakdown when continuity of care is designed into the next evacuation, funded, staffed, and exercised. It has learned from communication failure when authority, interoperability, and public messaging protocols are rewritten and then tested under compound strain. Learning shows itself in changed defaults.

Still, caution is needed. Not every recommendation should be implemented exactly as written. Some after-action reviews are drafted under political pressure, incomplete information, or narrow vantage points. Some recommendations conflict with one another. Some are too costly relative to the marginal gain they offer. Some overfit the last disaster rather than the broader pattern. A mature learning system therefore does not confuse implementation with obedience. It compares, prioritises, budgets, and adapts. But that necessary selectivity can itself become an excuse for drift if institutions never make the trade-offs explicit. "Complexity" is one of the great burial grounds of lessons.

This is why follow-through is the right metric. Follow-through does not require perfection or total implementation. It requires evidence that lessons altered the system in some durable way. That evidence may lie in procurement changes, staffing protection, revised contracts, capital planning, public dashboards for implementation, empowered local reporting, mutual aid redesign, or exercises that deliberately pressure the weak points exposed by real events. Follow-through is measurable because it leaves administrative traces. Rhetoric leaves transcripts. The difference matters.

The strategic significance becomes especially sharp under compounding shocks because these events repeatedly expose not new hazards alone, but old administrative habits under new pressure. If the state cannot learn across sectors, then each compound emergency will appear more surprising than it really is. If it cannot protect preventive spending against the politics of visibility, then visible response will continue to crowd out the quieter work that might reduce the next crisis. If it cannot create reporting cultures that are candid without becoming purely punitive, then the information needed for adaptation will arrive late and filtered. Institutional learning is therefore not simply one good thing among others. It is one of the capacities that determines whether state weakness hardens or recedes after shock.

The file is written after every serious emergency. The harder question is whether it remains a record or becomes a route. When the next review identifies the same shelter weakness, the same fuel dependency, the same staffing bottleneck, the same communications seam, the same mutually assured optimism about aid arriving from elsewhere, what should count as proof that a government has learned: a new phrase in the summary, or a budget, contract, roster, and authority structure that no longer assume the lesson belongs to history rather than to the next storm?

Chapter 7

Budget Cycles and Catastrophic Uncertainty

What looks prudent in a single budget year can become reckless across a generation. That is the paradox at the heart of public finance under climatic stress. Governments are asked to justify spending against visible needs, auditable benefits, and electoral time. Risks that mature slowly, arrive unevenly, or reveal their value only when disaster does not happen are therefore placed at a disadvantage before any argument begins. A temporary repair can be counted, photographed, and defended. A redesign that prevents a later cascade often appears expensive, speculative, or politically indulgent. The patch holds.

That patch is more than a maintenance choice. It is a fiscal habit. Public budgeting is usually organised to reward near-term balance, administrative clarity, and the appearance of efficiency. Those are not trivial goods. They protect against waste, patronage, and magical thinking. Yet they also create a recurring blind spot. The longer the horizon of the threat, and the less visible the avoided harm, the easier it becomes to defer action while still sounding responsible. Deferral then acquires the dignity of prudence. A minister can say the asset is safe enough for now, the review is ongoing, the redesign will be reconsidered next cycle, the reserve can be postponed, the contract can be extended one more year. The patch holds again, and the underlying exposure quietly grows.

This is why political time matters so much. Climate-linked risk does not arrive on the same calendar as mandates, fiscal years, or spending reviews. A coastal depot, remote housing scheme, or transport corridor may be planned on the assumption of decades of use. The politician approving or delaying it may answer to a far shorter rhythm. Within that shorter rhythm, deferral can feel entirely rational. It reduces immediate conflict, avoids visible tax or debt implications, and leaves the harder choice for a successor. At the level of the whole system, however, repeated deferral becomes its own strategic exposure. The state has not merely postponed expense. It has chosen to keep betting on stability while narrowing its future room to adjust.

That narrowing matters because public finance is not only about what a government can spend. It is about what options remain open when stress arrives. A resilient fiscal system is not simply a generous one. It is one that retains contingency capacity: reserves that can absorb shock, procurement systems that can surge without paralysis, staffing rules that do not treat all slack as waste, and balance sheets that can support emergency action without turning every disruption into a political panic about affordability. The hard distinction here is between resilience and austerity misdescribed as discipline. A system shaved too close to the line may look efficient in calm periods and prove incapable when several demands arrive together. The patch holds until the next stress tests what "efficient" really meant.

Infrastructure makes this especially plain because capital projects are never just projects. They are promises about future conditions. To build a depot, runway, housing estate, power connection, seawall, or hospital is to assume something about the world in which that asset will operate, the maintenance regime that will support it, and the public obligations that will follow from its existence. Once built, such assets can lock governments into patterns of service, liability, and political expectation that are difficult to reverse. This does not argue against building. It argues for candour. Uncertainty changes what responsible building means. It raises the value of modularity, reversibility, and exit options, not because caution is noble in itself, but because fixed commitments made under unstable assumptions can become expensive traps disguised as development.

The stakes extend beyond technical planning and into legitimacy. Publics rarely evaluate budgets through discount rates or option value. They evaluate them through narrative. Was money spent on something real or on a speculative fear? Was a reserve prudent or a sign that officials expected failure? Was a redesign wisdom or overreaction? Was under-spending restraint or neglect? These questions matter because legitimacy is itself a fiscal resource. A government that cannot explain why preventive spending is serious spending will struggle to defend it. A government that repeatedly underfunds quiet prevention while funding visible response after each failure teaches the public a different lesson: that urgency, not prudence, is what unlocks money. That lesson becomes expensive.

The path ahead follows this fiscal logic from immediate habit to long commitment. It begins with the budget tools and political incentives that make deferral appear rational even when the system grows more exposed, then moves into the quieter capacities - reserves, surge procurement, staffing flexibility, financial slack - that determine whether a state can absorb shocks without administrative panic. From there the argument turns to infrastructure as an intergenerational promise rather than a one-off expenditure, before arriving at the stories governments tell about risk and why those stories help decide which budgets look legitimate before crisis and which only after it.

What binds these themes is not a dispute about whether states should spend more or less in the abstract. It is a more exacting question about timing, optionality, and candour. Public finance becomes strategically weak when it confuses near-term tidiness with long-term control, when it counts visible outputs and discounts avoided losses, when it treats flexibility as inefficiency, and when it imagines that a later government will inherit the same range of choices if action is delayed now. Catastrophic uncertainty punishes that imagination. It turns ordinary budget routines into vehicles for compounding exposure.

This is also why later debates over governance and communication are already present here in embryo. A reserve fund, a redesigned standard, a modular capital plan, or a more generous maintenance regime will always need a story that makes it politically bearable. Without that story, prudent spending can be mocked as waste and repeated patching can masquerade as realism. Fiscal systems do not merely allocate resources. They encode what a polity is willing to treat as foreseeable, tolerable, and worth paying to avoid. Those judgments are never only technical. They are moral and strategic at once.

The patch remains the image worth keeping. Not because temporary repairs are always wrong, but because they so neatly reveal the chapter's governing tension. A patch can be sensible, merciful, and efficient in one season. It can also become a ritual through which a state proves, year after year, that it has chosen political ease over strategic adjustment. The harder question, then, is not whether governments can afford to plan for catastrophic uncertainty. It is how long they can afford to keep mistaking the patch for a budget philosophy before the bill for that philosophy arrives all at once.

Discounting, deferral, and political time

In May 2020, after the diesel tank collapse near Norilsk in northern Siberia, the public saw a stark environmental disaster and a brutal industrial failure.

The operator was NTEK. The place was the Russian Arctic. The date is easy to verify. But there was another story inside the same event, quieter and more familiar to public finance: a long-lived asset had been maintained inside assumptions that proved cheaper to preserve year by year than to challenge in time. The spill looked sudden. The budgeting logic behind it was not. The patch had held.

That image matters because public budgeting often fails not through negligence in the crude sense, but through repeated acts of plausible postponement. A government does not usually declare that long-horizon risk should be ignored. It reviews, defers, studies, phases, pilots, and extends. It chooses temporary stabilisation over redesign, emergency appropriation over capital re-siting, short-term repair over deeper adaptation, another maintenance cycle over the politically difficult admission that the asset was built for conditions no longer likely to hold. Each decision can be defended within the fiscal year or electoral term in which it is made. Taken together, they form a pattern of structural exposure. The patch had held again.

This is where discounting enters the argument, not merely as an economic technique but as a political habit. Standard budget tools are designed to compare costs and benefits across time. In principle, that is sensible. Governments should not spend indiscriminately on distant dangers while present needs go unmet. Yet the same tools tend to underweight harms that mature slowly, arrive unevenly, or produce benefits chiefly in the form of disasters that do not happen. Prevention is notoriously hard to defend because its success appears as absence. A road that does not fail, a runway that does not close, a depot that remains usable, a hospital that keeps power during regional stress - these outcomes are politically quiet. The ribbon-cutting belongs elsewhere. The avoided emergency leaves little theatre behind it.

That theatre gap is one reason deferral feels rational. Within a single mandate, postponing a costly redesign can preserve fiscal space, avoid conflict with voters or finance ministries, and maintain the appearance of prudence. A minister can say the asset remains serviceable, the review continues, interim repairs are underway, and stronger action will be considered once evidence sharpens or the next spending cycle opens. These are not always dishonest claims. Often they reflect real uncertainty and real constraints. The problem lies in aggregation. When several administrations, departments, or agencies make the same individually rational decision in sequence, the system as a whole begins to behave irrationally. It spends repeatedly to preserve a weakening assumption rather than once to alter it. Political time compresses responsibility and stretches exposure.

That compression is the deeper meaning of political time. Hazards unfold over physical time. Assets age over engineering time. Public debt is managed over fiscal time. Electoral mandates, media attention, and ministerial accountability often operate over much shorter intervals. These clocks rarely align. Climate-linked infrastructure risk is especially sensitive to that misalignment because it affects long-lived commitments whose deterioration may be visible in fragments long before the full consequences become politically undeniable. A retaining wall settles slightly. A runway needs another temporary correction. A road section is regraded one more time. A fuel depot receives another short-term stabilisation. The individual decisions remain explainable. The cumulative trajectory is harder to defend because no single office, term, or budget fully owns it.

A useful example can be found in repeated temporary repairs to transport infrastructure in difficult environments, especially where permafrost thaw, drainage change, or recurrent flooding undermine older design assumptions. A road segment or runway can often be patched successfully for a season or two. The patch is cheaper than redesign. It is quicker to approve. It satisfies the immediate complaint. It allows officials to say the issue has been addressed. But repeated temporary repairs substitute for redesign most often when the budget system rewards visible maintenance output over harder strategic admission. Eventually the patch stops being a bridge to a better plan and becomes a fiscal doctrine in miniature: preserve current service at the lowest immediate political cost, even if the cost of preserving it rises with each round. The patch had held long enough to become policy.

This is not only a problem of technical underestimation. It is also a problem of benefit visibility. Suppose a government proposes moving a storage site, raising a road section, redesigning a shoreline facility, relocating housing from unstable ground, or building a more modular depot with exit options. The immediate cost is visible. The counterfactual benefit is harder to dramatise. Citizens and legislators see the expenditure. They do not see the crisis that fails to materialise ten years later, the insurance premium that does not rise as sharply, the evacuation that remains possible because the route still functions, or the budget crisis that never takes shape because the asset does not require serial emergency repair. Where democratic politics depends heavily on visible returns, invisible benefits are always at a disadvantage unless they are narrated well and defended consistently.

This is why discounting becomes political even when it is presented as neutral arithmetic. The farther into the future a harm is pushed in budget models or political rhetoric, the easier it is to justify spending less now. But the issue is not only how future costs are formally valued. It is also how future responsibility is socially discounted. Voters may assume a later government will cope. Departments may assume a future review will settle the matter. Officials may believe that stronger action can wait until after the next report, election, audit, or incident. The result is not ignorance but managed postponement. What appears as prudence in one file becomes accumulation in the public realm. Structural exposure is often nothing more mysterious than delayed recognition institutionalised.

The example of Newtok, Alaska, is useful here, not because it mirrors every case of infrastructure vulnerability, but because it reveals how long systems can live in the gap between recognised risk and fully funded, politically settled response. The community's exposure to erosion and unstable conditions was not a secret suddenly discovered by science or government. It was known, argued over, planned around, and deferred through multiple institutional and fiscal cycles. The specific legal and political complexities were substantial. So were the logistical and financial burdens. The relevance lies precisely there. Long-horizon risks do not defeat public systems because nobody sees them. They often defeat them because everyone sees them through a different clock. The state knows and still delays because no single moment seems quite decisive enough to overturn the spending logic of the moment at hand.

A second example lies closer to ordinary infrastructure maintenance and therefore perhaps closer to the real centre of the problem. After repeated flood or storm damage, local governments often face a choice between another cycle of temporary repair and a more expensive redesign or relocation. Repeated temporary repairs can be entirely rational for one season, especially if funds are scarce and the political pressure to restore service is immediate. Yet when this pattern repeats over years, the fiscal system begins to reveal what political time has concealed. Insurance costs rise. Emergency appropriations become normal. Maintenance budgets are cannibalised by recurring damage. Other services are delayed to protect the same vulnerable asset again. Credit conditions may tighten. What was framed as thrift becomes serial spending in defence of an increasingly fragile baseline. The patch had held, but the ledger had started to say otherwise.

A fair counter-example must be acknowledged. Not every case of deferral is irrational. Sometimes redesign genuinely should wait because uncertainty is too large, engineering options remain immature, population change is unclear, or the asset's strategic importance is already declining. Temporary repair can be wise bridge management rather than denial. Standard discounting tools can also protect against panic spending on highly speculative scenarios. The argument is not that all deferral is bad or all long-term prevention good. It is that conventional budget discipline often treats long-horizon risk as though postponement were costless until evidence becomes overwhelming. In systems already exposed to climate-linked instability, that threshold may arrive after the menu of sensible options has narrowed sharply.

This is why political time should be understood as a generator of exposure rather than merely a backdrop to it. The state does not only inherit hazards from the climate or the market. It manufactures vulnerability through repeated decisions to privilege short-term fiscal legibility over long-term strategic room. A road left in place one more cycle may constrain future relocation choices. A housing development approved on optimistic assumptions may create decades of service obligations in a risky zone. A depot sited for present convenience may become expensive to defend as surrounding conditions change. Each decision creates lock-in. None feels catastrophic at the moment of signature. Catastrophic uncertainty is often built from these ordinary commitments.

The moral and political difficulty is that governments are not wrong to care about present burdens. People need services now. Budgets are finite. Tax tolerance is real. The challenge is therefore not to replace present needs with future anxieties, but to resist the false comfort that present restraint always protects future stability. Sometimes it does the opposite. Cutting reserves, delaying redesign, preserving exposure through cheap repair, or discounting the value of optionality can produce a system that looks fiscally sober until the next compound shock reveals how little room was really left. Austerity and resilience are not opposites in every case, but neither are they synonyms. Much depends on whether a budget preserves the capacity to change course later or quietly uses it up.

This is where repeated temporary repairs become more than a technical detail. They are signals about what a polity is willing to admit. A government that patches repeatedly may be saying, in effect, that the deeper redesign is unaffordable, politically impossible, or insufficiently urgent for current voters. It may also be saying that future governments will have to live with the consequences. The choice is not always avoidable. But it should at least be named accurately. The danger lies in pretending that the patch is simply common sense while the larger exposure is left offstage. Public finance becomes strategically weak when it cannot describe its own postponements honestly.

Narrative plays a decisive role here. Preventive spending looks wasteful when governments cannot explain what it is buying beyond abstract safety. Repeated patching looks reasonable when failure is narrated as bad luck, exceptional weather, or unfortunate timing rather than the predictable outcome of deferred adaptation. Citizens are capable of understanding risk over time, but they rarely receive public language equal to that task. They are shown invoices for redesign and almost never the cumulative bill for deferral. They are asked to admire fiscal restraint in the present and to endure emergency spending in the future as though the two were unrelated. Political time thrives on that split.

This section therefore points towards a harder but more honest way of reading budgets. Instead of asking only whether a repair or redesign is affordable this year, one should ask what assumptions about future stability the decision requires, what options it preserves or closes, and whether the seeming prudence of deferral depends on someone later paying a larger price under worse conditions. That is the strategic question hidden inside discounting. Numbers matter, but so do the clocks behind them. One administration's restraint can become the next administration's trapped inheritance.

The patch had held at Norilsk until it did not, and countless smaller cases repeat the same fiscal logic without ever reaching global headlines. That is why the core question is not whether governments should stop using discounting or forbid temporary repairs. It is whether they can learn to recognise the point at which those tools stop helping them manage uncertainty and start teaching them to live inside a shrinking horizon - one budget cycle at a time, until the system's accumulated rationality finally looks like something else entirely.

Fiscal resilience and contingency capacity

In March 2020, governments across the world discovered that the question "what do we have in reserve?" was not abstract at all. The place could have been London, Ottawa, Rome, or Wellington; the pattern was recognisable everywhere. Public systems did not fail only where they lacked money in the narrow sense. They failed where they lacked room to move: staff who could be redeployed without hollowing out other services, contracts that could expand without months of delay, stocks that existed before panic buying, fiscal headroom that could absorb uncertainty without turning every emergency purchase into a political drama. The cupboard mattered.

That image of the cupboard matters because fiscal resilience is often misunderstood as a synonym for wealth, or else as a sentimental plea for "more resources". Neither is sufficient. A rich system can be brittle if its money is trapped in rigid categories, its hiring rules are slow, its procurement is optimised for peacetime neatness, and its reserves have been shaved away in the name of efficiency. A poorer system can sometimes absorb shock better than expected if it preserves practical optionality: emergency purchasing authority, flexible staffing, modest buffers, local discretion, and a treasury culture that can tolerate temporary disorder in pursuit of continuity. Fiscal resilience, then, is not simply how much a state has. It is how much it can still do when conditions turn against the plan. The cupboard matters again, because what matters is not only what sits on the shelf but whether anyone can reach it in time.

This is why contingency capacity deserves to be brought out of the shadows of budget talk. Reserves, surge procurement, staffing flexibility, emergency contracting, discretionary funds, warehoused supplies, and standby service arrangements are frequently treated as inefficiencies, leftovers, or temptations to waste. Under routine conditions they may indeed appear underused. That appearance is part of the political problem. Their value materialises chiefly when the emergency does not wait for the next appropriations cycle or the next competitive tender. The concept to name here is optionality under stress. It refers to the state's ability to preserve meaningful choices when information is incomplete, time is compressed, and several systems need support at once. The implication is severe: a government that has optimised away its options in calm periods may discover that it has preserved balance on paper by spending its freedom of action in advance.

The distinction between resilience and austerity therefore has to be drawn more carefully than public argument usually allows. Austerity is not simply low spending, and resilience is not simply high spending. The real difference lies in what the budget protects. A state may reduce costs while still preserving emergency authority, critical spare capacity, and the ability to reallocate quickly. Another may boast of discipline while hollowing out exactly the slack that compound crises require: reserve funds treated as idle cash, warehouse stock judged inefficient inventory, staffing pools reduced to bare minimum, maintenance crews replaced by contingent arrangements that work only when demand is not simultaneous across a region. To ask whether a budget is "tight" or "generous" is often the wrong question. The sharper question is what room remains when the ordinary script fails.

A clear example came during the early months of the COVID-19 pandemic, when even wealthy countries struggled not only with demand surges but with the conversion of fiscal intent into practical supply. Governments authorised spending quickly enough in some cases, yet still found themselves constrained by procurement bottlenecks, shallow inventories, thin supplier networks, rigid staffing systems, and poor information about what existed where. The lesson for this book is not about public health in isolation. It is about fiscal architecture. Money was necessary. It was not identical with capacity. A treasury can approve emergency funds at noon and still fail to produce ventilators, protective equipment, locum staff, or transport by nightfall if the intervening institutions have no surge mechanism. Optionality lives in those intervening institutions.

Climate-linked shocks expose the same architecture in different form. A flood, fire season, heat emergency, major evacuation, permafrost-related infrastructure failure, or prolonged utility disruption does not ask only whether a budget line exists. It asks whether reserve funds can be accessed without destructive delay, whether procurement rules allow temporary deviation from routine sequencing, whether relevant staff can be reassigned without collapsing another service, whether framework contracts include realistic surge clauses, whether supplier concentration has created hidden bottlenecks, and whether local managers possess enough fiscal discretion to solve a problem before it grows into an appeal for ministerial intervention. A budget can therefore appear prudent while being operationally slow. That slowness is often mistaken for scarcity. Sometimes it is scarcity. Often it is rigidity disguised as order.

This is where the language of efficiency begins to do real damage. Efficiency reforms have obvious appeal. They promise leaner administration, lower carrying costs, fewer idle assets, better use of data, and clearer accountability. Much of that can be desirable. The danger comes when efficiency is defined so narrowly that all visible slack becomes a target. A spare warehouse is questioned because occupancy rates are low. A reserve contract is attacked because it is rarely triggered. A staffing cushion is reduced because utilisation figures look weak. A local emergency fund is centralised because unspent balances appear untidy. Procurement is streamlined around fewer suppliers because scale promises savings. Each decision can make sense on its own terms. Together they can remove the very redundancies and options that make shock absorption possible. The cupboard looks disciplined just before it looks empty.

The same logic applies to staffing. Financial discussions often treat labour as a recurrent cost to be contained. That view is not irrational; labour is expensive, and public institutions should be careful. Yet crises are governed through people before they are governed through spreadsheets. A reserve of trained staff, cross-trained personnel, locum capacity, mutual support agreements grounded in real availability, and administrative posts that can expand or contract quickly under stress all form part of fiscal resilience even though they may look like overhead in calm periods. If a water utility has no spare technical staff, if a hospital has no roster flexibility, if an emergency management office has no procurement officer available when contracts must be altered quickly, money alone will not create capacity on demand. The cupboard includes people.

This is why surge procurement deserves more scrutiny than it usually receives. Governments often assume that when crisis hits they can "buy what is needed" provided the treasury approves. In reality, surge procurement is a specialised capability with its own dependencies. It depends on market depth, legal authority, pre-existing frameworks, trusted suppliers, warehousing, transport, technical specifications, payment systems, and staff who know how to act quickly without collapsing into either paralysis or fraud risk. A state that has centralised procurement aggressively, consolidated suppliers too tightly, or cut purchasing expertise in the name of streamlining may discover that its emergency buying power is far weaker than budget documents imply. Emergency purchasing is not just fiscal permission. It is an institution.

A useful counter-example lies in systems that maintain reserves or contingency financing precisely because they expect volatility. Some resource-dependent jurisdictions, and some states with stronger disaster traditions, have built rainy-day funds, contingency mechanisms, or emergency purchasing protocols that can be activated quickly when shocks arrive. These arrangements vary widely in design and quality, and no simple model should be romanticised. But they do reveal the principle. Fiscal resilience improves when institutions accept that some unspent money, some underused contracts, some spare stock, and some flexible staffing will appear inefficient until the day they are suddenly not. This is not an argument for limitless reserves or permanent overcapacity. It is an argument against the fantasy that all visible slack is waste.

Still, caution is necessary. Contingency capacity can be designed badly. Reserve funds can become symbolic pots raided for unrelated purposes or so politically protected that they are not used when they should be. Emergency procurement can invite poor value, patronage, or weak oversight if rules are loosened without discipline. Staffing flexibility can be abused, leading to burnout or hidden precarity. Warehoused supplies can expire, degrade, or create false confidence if logistics are not maintained. A serious account must admit these risks. The defence of optionality is not a defence of disorder. It is a defence of capabilities that require active stewardship. Slack without governance becomes waste. Governance without slack becomes brittle.

That brittleness is especially dangerous under catastrophic uncertainty because uncertainty changes the value of room to manoeuvre. A government facing a known and stable risk can optimise more tightly around expected demand. A government facing uncertain timing, compounding hazards, and unevenly distributed shocks needs more capacity to absorb surprise without immediate political or administrative breakdown. Fiscal resilience in such a setting is less about predicting perfectly and more about preserving reversibility. Can a budget absorb the first shock without forcing immediate cuts elsewhere? Can managers procure urgently without losing all control over cost and quality? Can staff be moved without making another critical service fail? Can a reserve be used before the crisis becomes headline-worthy? These questions render resilience legible in financial terms without reducing it to one accounting ratio or ideological preference.

Public trust is bound up with this more closely than budget debate usually admits. Citizens rarely inspect contingency funds or supplier frameworks directly. They do notice when public systems can respond without visible improvisation. They notice when shelters open with enough staff, when backup generators actually have fuel, when road repairs begin before bureaucratic confusion fills the news, when hospitals can surge without immediate collapse elsewhere, and when governments explain spending decisions in ways that match what people experience on the ground. Trust, in fiscal terms, is partly the belief that there is still something in the cupboard and that those in charge can open it without fumbling for the key. Once repeated emergencies teach the public that every response begins with frantic requests for emergency appropriations, waivers, and mutual aid that may not arrive, the credibility of the system changes.

This is why contingency capacity should be evaluated less by its static size than by its convertibility into action. A reserve fund that cannot be drawn quickly is less resilient than a smaller one with clear triggers and delegated authority. A supplier framework that exists on paper but has never been stress-tested may be weaker than a narrower arrangement with real surge performance. A staff pool that looks modest may outperform a larger nominal workforce if cross-training, overtime rules, and local discretion are more realistic. Fiscal resilience is therefore best measured through friction. How many steps stand between recognition of need and legally authorised action? How many between authorised action and delivery? Where does the process depend on single officials, single firms, single depots, or single interpretations of the rules? The answers tell us more than the headline size of the reserve.

The language of "fiscal discipline" often obscures these distinctions because it rewards appearances that are politically legible and morally flattering. A lean system looks responsible until a shock reveals that its tidiness was purchased by sacrificing options. A buffered system can look indulgent until the same shock shows that it can absorb pressure without panic. This inversion is one reason climate-related preparedness is so often underfunded. The capacities that matter most are frequently those least visible in success and most awkward to defend in normal years. They are hard to inaugurate, easy to mock, and essential when the ordinary timetable fails. Budget politics, left to itself, tends to select against them.

A second named example reinforces the point from another angle. During repeated wildfire seasons in jurisdictions such as California and parts of Australia, public systems have had to cope not only with suppression costs but with evacuation support, shelter operations, health impacts, infrastructure repair, and long recovery tails. The financial challenge has not been simply "spend more". It has been whether reserve funds, emergency contracting, interagency payment systems, and staff capacity could expand quickly enough without stripping other functions bare. Different jurisdictions have handled this with varying degrees of success, but the common lesson is clear. The true test of fiscal resilience arrives not when a treasury announces support but when support reaches the field before the system starts cannibalising itself.

This argument prepares the ground for the chapters that follow because it turns budget talk away from ideology and towards structure. One need not prescribe a single fiscal doctrine to see that some forms of "efficiency" reduce optionality and some forms of restraint preserve it. One need not romanticise redundancy to see that a state without reserves, surge capacity, or flexible staffing is not disciplined in any meaningful strategic sense. It is exposed. The difference matters because climate-linked disruption punishes systems that cannot convert recognition into action without a fresh political struggle each time.

The cupboard mattered in March 2020, and it matters whenever fires, floods, outages, evacuations, or infrastructure failures demand more than the ordinary script can provide. The hard question is not whether governments should keep more money or more things "just in case" in some vague sense. It is whether they can learn to see reserves, surge procurement, and staffing flexibility as financial forms of sovereignty - capacities that preserve choice under stress - or whether they will continue to treat them as inefficiencies until the next shock proves that the real extravagance was believing the cupboard could stay this bare and still count as preparedness.

Infrastructure as intergenerational commitment

In January 1963, Italy completed the Vajont Dam in the Alps, a project that embodied the confidence of an era in engineering scale and national development. The place was northern Italy. The date is easy to verify. The dam itself was not destroyed by the later disaster; a massive landslide into the reservoir drove a wave over the structure and devastated communities downstream. The case is often remembered as a catastrophe of geology, design, and warning ignored. It should also be remembered as a lesson in what infrastructure really is. A capital project is never only concrete, steel, and cost. It is a wager about future conditions and about the institutions that will inherit the consequences of that wager. The site remembers.

That is the right place to begin because public discussion often treats infrastructure as a matter of provision rather than commitment. A road is built to solve access. A depot is built to secure supply. Housing is built to meet demand. A port is built to move goods. All true, and all incomplete. To build is also to bind. It creates obligations of maintenance, service, liability, protection, and political defence that outlast the administration that approved the project and often outlast the assumptions under which it was justified. This is what makes infrastructure intergenerational in the strongest sense. It is a promise, written into land and budget, that future governments and future residents will live with conditions sufficiently close to those imagined at the moment of construction. The site remembers again, often more accurately than the speeches do.

Under climate uncertainty, that promise becomes more difficult and more strategic. The challenge is not that governments should stop building because the future cannot be known perfectly. States cannot govern through suspension. People need homes, depots, runways, roads, clinics, and ports now. The challenge is that uncertainty changes what responsible building means. A project justified under assumptions of stable shoreline, dependable freeze-thaw cycles, manageable flood recurrence, or enduring settlement patterns may become a source of lock-in if those assumptions weaken. The concept to name here is inherited exposure. It matters because the cost of being wrong about future conditions is not merely technical retrofitting. It can be decades of political, fiscal, and moral obligation to defend an asset or settlement pattern that no longer fits the terrain.

Lock-in is the first term that requires sharper treatment. It is often used vaguely, as a synonym for path dependence or policy inertia. Here it should mean something more concrete. A piece of infrastructure locks in a pattern of public and private behaviour that becomes expensive to reverse. A fuel depot placed in a vulnerable location is not simply a storage decision. It shapes transport routes, emergency planning, insurance exposure, and future arguments about whether relocation is affordable. Housing built in a risk-prone area does not only shelter people in the present. It creates future claims on drainage, roads, schools, utilities, rescue, compensation, and political attention. Lock-in is therefore not a side effect. It is one of the main products of capital spending. The implication is severe: a project can look efficient at approval and strategically expensive in use because it creates obligations that later governments cannot easily refuse.

This is where uncertainty should change how planners think, not by paralysing action but by widening the meaning of prudence. Under relatively stable conditions, a state may rationally optimise around durable siting, long asset life, and integrated service patterns. Under less stable conditions, prudence may require attention to modularity, reversibility, and exit options. These terms are often reduced to design jargon. They deserve political reading. Modularity means an asset or settlement pattern can be adapted, expanded, relocated, or partly retired without catastrophic disruption to the whole system. Reversibility means governments can change course without incurring prohibitive political or fiscal penalties. Exit options mean the state has not built itself into a moral corner where retreat becomes unthinkable even after the original assumptions fail. These are not luxuries for nervous planners. They are forms of strategic freedom purchased in advance.

Show the argument in a familiar setting and its force becomes clearer. A government decides where to place a new bulk fuel depot serving remote communities. One site appears cheaper because it aligns with existing transport links and requires less immediate capital work. Another site is more expensive upfront but less exposed to future shoreline instability, ground movement, or repeated access disruption. A narrow budget logic may favour the cheaper option, especially if the visible savings fall inside the current mandate while the risk horizon stretches well beyond it. Yet the depot, once built, will anchor routes, insurance assumptions, emergency response plans, and perhaps settlement stability itself. If conditions deteriorate, later governments may discover that the original saving was small compared with the cost of defending, relocating, or repeatedly repairing the now indispensable asset. The issue is not simply engineering miscalculation. It is political time embedded in concrete.

Housing shows the same pattern in a more intimate register. Build homes in a location that appears serviceable now but depends on fragile drainage, unstable ground, exposed access, or expensive protective works, and the decision will not remain a one-off act of provision. It will become a decades-long commitment to roads, utilities, insurance pressure, social service continuity, school transport, and perhaps eventual compensation or relocation assistance. Residents will rightly treat those homes not as provisional experiments but as evidence that the state judged the place worthy of durable life. That judgement, once embodied in mortgages, schools, clinics, and neighbourhood expectations, becomes hard to revise. A planning error becomes a political obligation.

Newtok, Alaska, remains one of the clearest examples of this general problem, even though its circumstances are specific and should not be treated as universal. The community's long struggle with erosion and relocation has shown how difficult it is to unwind inherited exposure once housing, public facilities, and legal responsibilities have accumulated in a location under mounting stress. The point is not that every vulnerable settlement can or should move, nor that all future housing decisions resemble Newtok. The point is that infrastructure and settlement choices generate obligations that later become expensive, contested, and morally charged to reverse. Building is easy to announce. Unbuilding is not. The site remembers, and so do the people asked to leave it.

This is why infrastructure must be understood as a political promise about future conditions. Governments do not merely provide assets; they imply a view of the future stable enough to justify those assets. A road says this route will matter and can be kept open. A housing estate says this place can sustain life with ordinary public support. A depot says this node will remain practical enough to anchor supply. A port expansion says access will remain worth the defence and maintenance it requires. Such promises are never made in a single sentence, yet citizens and markets hear them clearly. If conditions later undermine them, the question becomes not only what failed physically, but what kind of judgement the state exercised when it made the promise in the first place.

A second example lies in the siting of coastal and northern storage or transport infrastructure that later requires repeated protection works. Governments and firms often defend these projects by pointing to sunk costs and current dependence: the asset exists, communities rely on it, moving it would be too expensive, therefore it must be defended. That logic can be entirely understandable. It can also conceal the strategic consequences of earlier choices that treated present convenience as if it were the same thing as future viability. Once enough dependent systems gather around the original site, every later debate becomes harder. The state is no longer choosing where best to place a facility under uncertainty. It is choosing whether to admit that the earlier placement created a trap. Lock-in turns a planning issue into a legitimacy issue.

Still, caution is necessary. The argument against naive stability bets is not an argument for endless hesitation or fetishised flexibility. Not every project should be modular. Not every asset needs a graceful exit path. In some cases, permanence is both necessary and sensible. Communities cannot be governed through provisionality alone. There are real costs to overbuilding for uncertainty, just as there are real costs to underestimating it. Modular systems can be more expensive, more administratively complex, or less efficient in normal conditions. Exit options can become excuses for political non-commitment, leaving citizens to feel that the state has built without believing. These are serious limitations. The point is not that uncertainty nullifies the value of durable infrastructure. It is that durability now needs a richer meaning than simple longevity at one fixed site.

A useful counter-example comes from infrastructure systems deliberately designed to evolve in stages. Some ports, transport corridors, and public facilities are built with phased expansion, movable components, or adaptation pathways that allow governments to add, retreat, or reconfigure as information improves. This does not abolish risk. It does show that capital planning need not be trapped between reckless permanence and timid indecision. The lesson worth taking is not architectural fashion but institutional humility: a state can commit seriously while preserving some ability to change course if the conditions underlying the original commitment begin to fail. Humility, here, is not weakness. It is a form of strategic respect for the fact that future governments must govern within the decisions made today.

This becomes a fiscal issue as quickly as it is a planning one. A rigidly sited or highly integrated asset may look cheaper at approval because it minimises immediate capital cost. A more modular design, higher site, or dispersed system may appear unnecessarily expensive. But immediate capital economy is not the only relevant measure. One must also ask what future budgets are being asked to underwrite if the first decision turns out badly. Will later governments face serial repair, higher insurance, protective works, legal disputes, community anger, and shrinking room to relocate because too much has already been built around the asset? If so, the original "saving" may have been achieved by pushing contingent cost onto future administrations and future taxpayers. Public finance often records this only belatedly. The site remembers faster than the ledger.

Legitimacy enters because people do not experience infrastructure as a portfolio of options. They experience it as a statement about belonging. When the state builds homes, roads, depots, ports, schools, or clinics, it signals that life in that place is meant to continue under the state's protection. To discover later that the same state now speaks the language of managed retreat, loss adjustment, or difficult trade-offs is not merely to encounter a revised technical assessment. It is to encounter a revision of political meaning. This is one reason infrastructure debates become so emotionally charged under climate stress. The physical asset is never only physical. It is evidence that the future once felt more stable, or at least more governable, than it now appears.

That is also why planners should ask a harder set of questions before locking in exposure. Not only: can this be built? Not only: does it meet current need? But also: what assumptions about future conditions make it sensible, what obligations will it create if those assumptions weaken, how hard will it be to adapt or retreat later, and who will bear the political and financial cost if the promise embedded in the asset turns out to have been overconfident? These are strategic questions because they concern room for manoeuvre across generations, not merely present utility.

The brief example of siting decisions for depots and housing captures the wider argument because both asset types gather dependencies quickly.

A depot brings supply lines, emergency planning, transport rhythms, and regulatory commitments. Housing brings schools, clinics, drainage, roads, tax expectations, community memory, and the moral claim that ordinary life should be possible there. Once these patterns crystallise, later adaptation becomes more expensive materially and politically. The site remembers through systems, not through soil alone.

What follows from this is not a counsel of abstention but a demand for cleaner candour. States will continue to build, and should. But if infrastructure is an intergenerational commitment, then the real question is not whether to commit. It is what kind of commitment is being made: one that quietly assumes stability because the current budget prefers it, or one that acknowledges uncertainty early enough to preserve some freedom for those who will inherit the site, the promise, and the bill.

Public narratives and budget legitimacy

In November 2012, after Hurricane Sandy struck the eastern United States, arguments over rebuilding, resilience spending, and public responsibility began almost as soon as floodwaters receded. The place was New York and New Jersey, among others. The date is easy to verify. What quickly became clear was that the political struggle was not only over money. It was over what the money meant. Was protective spending prudent foresight or extravagant overreaction? Were delayed repairs evidence of fiscal discipline or proof that leaders had mistaken neglect for normality? The invoice had a story attached.

That story matters because public budgets do not operate in a vacuum of technical reason. They move through narratives about danger, waste, prudence, fairness, and competence. A reserve fund, a redesigned drainage system, a relocated depot, a reinforced shoreline, or an upgraded emergency communications network will not be judged by engineering alone. It will also be judged by the language surrounding it. If governments cannot explain the risk that spending is meant to reduce, the expenditure can look ornamental, ideological, or indulgent. If governments repeatedly under-spend while speaking as though conditions remain ordinary, the absence of action acquires the dignity of common sense. The invoice, again, is never just a number.

This is why budget legitimacy should be treated as a strategic variable rather than a public-relations afterthought. Legitimacy in fiscal life means more than legal authority to tax and spend. It means the continuing ability to persuade citizens, legislatures, investors, and local institutions that the allocation of money reflects a credible understanding of risk and a defensible set of priorities. When that credibility weakens, prudent expenditure becomes harder to defend, delayed expenditure becomes easier to normalise, and later emergency spending can arrive in an atmosphere already primed for suspicion. The concept to name here is fiscal narration. Budgets are not only decisions about resources. They are stories about what kind of future the state considers plausible enough to prepare for.

Climate-linked risk intensifies this problem because the most sensible spending often produces benefits that are difficult to see directly. A redesigned culvert does not make headlines when it works. A reserve fuel contract is politically quiet if the emergency never requires it. Modular infrastructure, relocated storage, redundant communications links, or stronger inspection regimes usually demonstrate their value through the absence of later drama. Meanwhile, the costs are visible immediately, and often locally. Taxpayers see the outlay. Opponents see an opportunity. Auditors see a line item. The public rarely sees the avoided cascade. A government that wants legitimacy for such spending has to do more than produce a business case. It has to produce a public account of why the visible cost is less wasteful than the invisible exposure it is meant to reduce.

This is where communication failure can turn prudent spending into perceived waste. Suppose a government funds a more expensive depot site, a stronger evacuation route, or more generous contingency stocks without explaining clearly that the decision addresses reliability under growing uncertainty rather than prestige or bureaucratic caution. Critics can then frame the same expenditure as overbuilding, climate theatrics, or administrative indulgence. The spending may still be technically sound. Politically, however, it can be weakened by narrative emptiness. Citizens do not automatically infer the rationale behind resilience-oriented expenditure, especially when the benefit is a crisis that fails to occur. If the state has not prepared the public argument, the first visible evidence of success may be indistinguishable from unnecessary expense.

The reverse failure is even more common. Under-spending can be turned into normalcy by a narrative of routine continuity. Temporary repairs are framed as sensible maintenance. Deferred redesign becomes "further review". Reserve reduction becomes "efficiency". Thin staffing becomes "modernisation". Vulnerable infrastructure remains in service because leaders describe its exposure as manageable, exceptional, or still under assessment. None of these phrases is always dishonest. Taken together over time, they can produce a dangerous public impression that inaction is realism and preparedness is alarmism. Once that frame hardens, a government may find itself able to spend vast sums after a visible failure while struggling to defend a fraction of that amount beforehand. The public has learned the wrong budget lesson.

This is not an abstract problem of political language. It shapes real fiscal outcomes. Legislatures hesitate when preventive spending has not been narratively grounded. Local governments struggle to win support for redesign if the public has been taught to see current conditions as baseline normal. Treasury officials find it easier to approve emergency repair than anticipatory adaptation because emergency repair fits a familiar and morally uncomplicated story: something happened, now we must act. Adaptation asks for a harder kind of legitimacy. It asks citizens to accept expenditure on the strength of a future not yet fully visible and on harms that may remain unevenly distributed or probabilistic. If that legitimacy is not cultivated, the budget system will often default to repair-after-failure because failure is easier to explain than foresight.

A named example from Australia illustrates the point. After repeated severe bushfire seasons and flood events, public and political argument has often turned not only on the adequacy of response but on whether earlier mitigation, land management, infrastructure hardening, and preparedness spending had been taken seriously enough. These debates are rarely settled by technical evidence alone. They depend on narratives about responsibility, memory, and blame. Preventive spending can be praised in retrospect when disaster strikes nearby, and mocked as excessive when the season appears calmer than feared. The relevance here is not to choose sides in any one national argument. It is to show that funding choices become politically durable only when they are attached to a persuasive account of risk that survives both action and apparent false alarm.

A second example can be found after Hurricane Sandy. Protective investments, rebuilding standards, and infrastructure adaptation were debated not simply as engineering questions but as narratives of public purpose. Was it sensible to rebuild stronger, more selectively, or elsewhere? Was expensive protection an act of stewardship or a refusal to face changing exposure? Was delaying redesign prudent caution or a way of avoiding politically difficult truths? These questions were fiscal in form and narrative in content. They matter here because they show that public spending under climate stress is never judged solely by what it buys. It is judged by what story it appears to affirm about the future and about who deserves protection on what terms.

This is why legitimacy can properly be called a fiscal resource. Like other resources, it can be built, spent, and depleted. A government that repeatedly explains risk honestly, spends in ways that match its explanations, and revises its position candidly when evidence changes can build a reserve of trust that makes prudent but initially unpopular spending easier to sustain. A government that alternates between minimising risk and demanding emergency spending, or between theatrical warning and underfunded follow-through, draws down that reserve. Eventually even sensible expenditure encounters scepticism because the surrounding narrative has been damaged. Legitimacy depletion is not merely a matter of public mood. It appears in legislative resistance, audit hostility, municipal frustration, insurance retreat, and the growing difficulty of converting technical necessity into political permission.

The role of visibility is crucial here. Publics are more likely to accept visible response than invisible prevention, visible damage than invisible fragility, visible ribbon-cutting than quiet maintenance, visible emergency authority than invisible reserve capacity. These are not irrational preferences. They are consequences of how democratic attention works. Governments therefore face a genuine challenge. They must make invisible benefits narratable without overselling certainty, and make under-appreciated risks legible without sounding either apocalyptic or self-serving. This is where many fail. They either hide behind technical language that means little outside the department, or they reach for inflated rhetoric that later weakens credibility when the event does not unfold as dramatised. Fiscal legitimacy erodes under both habits.

This problem becomes sharper where risk is distributed unevenly. A coastal reinforcement project, a northern runway redesign, a resilient water system, or a depot relocation may benefit one region more than another, at least visibly. Citizens elsewhere may ask why their money is being spent there and not here. The answer cannot be a mere gesture to expert assessment if the broader public has not been given a framework for understanding systemic exposure, interdependence, and the cost of later emergency spending. Without such a framework, targeted prudence can be narrated as favouritism, and universal inaction can be narrated as fairness. This is one of the ways legitimacy is depleted: by forcing specialised spending through a public language too thin to carry its rationale.

Still, caution is necessary. Not every expensive proposal branded as resilience deserves legitimacy simply because it invokes climate risk. Governments can waste money, overbuild, choose symbolic projects over useful ones, or fund visible adaptation that does less than quieter alternatives would have done. Public scepticism is not always ignorance. Sometimes it is a healthy demand for better reasoning, clearer evidence, or more proportionate design. The answer is not to moralise resistance away. It is to improve the quality of the fiscal narrative and the traceability of the spending. A legitimacy-rich system is not one in which the public nods along automatically. It is one in which disagreement occurs within a framework that still accepts the seriousness of the risk and the good-faith burden of preparing for it.

That framework depends on repeated correspondence between words and action. A government that talks about climate resilience while repeatedly cutting maintenance, centralising emergency discretion without funding local capacity, or leaving known vulnerabilities in place will not retain fiscal credibility for long. Citizens and local institutions watch for mismatch. They notice when the same vulnerable road is repaired every year under a new slogan. They notice when contingency funds are announced and then quietly raided. They notice when a minister warns of future danger while dismissing current preventive expenditure as unaffordable. Such contradictions teach the public that risk language is opportunistic rather than guiding. Once that lesson settles in, even genuine prudence becomes harder to sell.

This is why public narratives should be understood not as packaging but as part of the budget mechanism itself. They decide which expenditures feel legitimate before results are visible. They shape whether prevention can survive a year without disaster and whether visible damage is interpreted as bad luck or accumulated neglect. They influence whether local officials ask for stronger protection early or wait for crisis to make the claim politically legible. They affect whether markets and insurers read a jurisdiction as candid and serious or evasive and reactive. Narrative, in this sense, is a fiscal instrument. Not because words replace money, but because words determine how much political room money has to move before the catastrophe.

A useful counter-example lies in jurisdictions or agencies that have managed to explain precautionary expenditure in plain public terms tied to specific service continuity. Where authorities can say, credibly and repeatedly, that the spending protects drinking water, keeps hospitals operating, secures the evacuation route, preserves communications, or prevents repeated repair waste, legitimacy is easier to sustain. The explanation does not need to be florid. It needs to be concrete. It must link present cost to a future that citizens can recognise as part of ordinary life rather than as abstract catastrophe modelling. This is not spin. It is democratic translation of technical risk into shared political understanding.

The deeper issue is that under climate-linked uncertainty, a government may lose budget legitimacy long before it loses fiscal solvency. It may still be able to borrow, tax, or reallocate. What it can no longer do easily is persuade people that prudent long-horizon spending is anything other than elite excess, or that repeated under-investment is anything other than ordinary common sense. Once that threshold is crossed, the public finance system becomes reactive by default. It spends big under visible duress and trims quietly in the preventive years between crises. Structural exposure then deepens while the narrative continues to flatter restraint.

The invoice had a story attached after Sandy, and after many other disasters besides. That is the point to keep in view before turning later to risk communication directly. A budget is not legitimised simply because it balances, nor delegitimised simply because it spends. Its authority depends on whether the story surrounding it helps citizens distinguish prudence from waste before the crisis, not only after it. The harder question, then, is not whether governments can afford resilience spending. It is whether they can still tell the truth about risk clearly enough that prudent spending remains politically possible before the next emergency turns all the missing expenditures into a much more expensive story.

Chapter 8

Migration Pressure and State Capacity

What happens when the safest available choice for a household becomes a test of administrative endurance for the place that receives it? That question is more useful than the louder arguments that usually dominate public discussion. People move for reasons that are rarely reducible to one cause and never neatly sorted into moral boxes. They move because work fails, crops fail, housing becomes untenable, violence grows nearer, water becomes harder to secure, schools and clinics stop being dependable, or because one season too many has taught them that waiting is no longer the prudent choice. Movement, in that sense, is not an aberration. It is one of the oldest forms of adaptation. The door opens.

Yet the same movement that lowers risk for one household can raise strain for another set of institutions. A city receives families more quickly than its housing stock can absorb. Schools add pupils before classrooms, transport, or language support are ready. Clinics see longer waits. Informal landlords divide rooms. Shelters built for short stays become semi-permanent. Border systems designed for one legal category meet several at once. Local officials ask for help from national ministries that are themselves operating through older assumptions about pace and scale. The door opens again, and the question shifts from movement itself to the capacity of public systems to remain orderly, humane, and credible under pressure.

This is where the chapter takes its bearings. Migration is neither a threat by definition nor a policy failure by default. It is a social fact that can reduce harm, preserve dignity, and create new forms of economic and civic life. It can also expose the thinness of housing markets, school systems, health services, registration processes, and local finance. The difficulty lies in the overlap. Public debate often reaches too quickly for grand language - invasion, chaos, crisis, burden, inevitability. Much of that language obscures more than it clarifies. The harder task is to examine how mobility interacts with state capacity, and how administrative systems can become stressed without turning people on the move into a threat category.

That distinction matters because the real break point is usually procedural before it is ideological. A border post must sort mixed flows under uncertain conditions. A municipal authority must find shelter space while keeping public health systems functional. A school system must absorb children whose arrival does not match the calendar or the staffing model. A housing market already stretched for existing residents becomes the first shock absorber for newcomers as well. These are not minor implementation details. They are the places where national rhetoric meets institutional throughput. When those places are underbuilt, delay and confusion are quickly moralised into political grievance. Capacity is judged through queues, wait times, and visible disorder long before it is judged through doctrine.

The stakes are therefore broader than migration policy in the narrow sense. They concern whether the state can translate humanitarian obligation, legal process, and practical service delivery into a coherent response that the public experiences as both fair and functional. If it cannot, several harms follow at once. People on the move face worse conditions, receiving communities face sharper strain, and political actors who profit from simplification gain ground by presenting administrative failure as proof of civilisational failure. That is why the chapter insists on evidence discipline. Overstatement is not only morally crude; it is analytically self-defeating. When every inflow is described as a security emergency, the real pathways of instability become harder to see.

Those pathways are more concrete than the public slogans suggest. Housing pressure is one. Registration and screening delay is another. Schooling and continuity of care become crucial very quickly when sudden displacement includes families, older people, or those with chronic illness. Informal settlement risk grows when formal absorption fails, and with it come pressures on drainage, sanitation, electricity access, and flood exposure that can turn a humanitarian challenge into a municipal one and then into a national political one. Cities become the practical sites where these pressures are sorted, deflected, or allowed to accumulate. Their capacity is therefore not local in consequence, even when it is local in administration.

The path ahead follows that administrative reality in sequence. It begins with mobility as a human adaptation that can protect households while loading receiving systems in ways that remain politically easy to misread.

It then turns to border governance, where humanitarian obligation and enforcement mandates are forced into daily coexistence through screening, sheltering, and coordination. From there the argument moves into the city, where absorption becomes a test of infrastructure, public health, and civic patience, before ending with a harder discipline around conflict claims, so that security analysis can trace pathways without turning climate and migration into a speculative morality play.

What joins these themes is the problem of classification. Institutions want clean categories because budgets, laws, and mandates depend upon them. Real movement is rarely so obliging. One family may be displaced by crop failure, debt, and local violence together. Another may move pre-emptively because the next season looks impossible to survive. A border officer, local registrar, school head, or clinic manager still has to act, even when the categories do not sit neatly. This is why migration under climatic stress is best understood as a planning domain. It asks whether institutions can process ambiguity without collapsing into either cruelty or paralysis.

That challenge also exposes a deeper truth about sovereignty. Sovereignty is often discussed as if it resided chiefly in exclusion, in the visible power to police a line or make a legal distinction. Those capacities matter. They are not the whole story. Public authority is also tested by reception, registration, triage, schooling, sanitation, housing, and the maintenance of public order without indiscriminate force. A state that can announce control but cannot manage arrivals, support municipalities, or sustain basic administrative fairness may still possess legal authority while losing practical legitimacy. The converse is also true: humane, orderly, well-coordinated absorption can strengthen rather than weaken the state's credibility because it shows that power is not merely declaratory.

The door remains the image worth keeping. A household opens it to leave risk behind. A city opens it reluctantly or generously, with systems strong enough to cope or already near their limit. A border opens and closes in legal time while human need keeps moving in social time. The harder question, then, is not whether mobility will continue; it will. It is whether states can build enough administrative depth that when the door opens under stress, what enters is not automatic disorder but a test of capacity they have already chosen to take seriously.

Mobility as adaptation and as stressor

In August 2023, authorities in the Northwest Territories ordered the evacuation of Yellowknife as wildfire threatened the city. The place was northern Canada. The date is easy to verify. The most immediate story was one of smoke, roads, aircraft, and urgency. The deeper story began as people moved. Families leaving danger were reducing risk for themselves in the only practical way available. At the same time, every place receiving them inherited new pressure: on housing, school spaces, clinic capacity, registration systems, transport, and public communication. The door opened.

That image is useful because migration is too often discussed as though it were either a threat in itself or an exceptional failure of order. Neither description is good enough. Human beings move when livelihoods weaken, when hazards become harder to absorb, when safety narrows, or when staying becomes a gamble with diminishing returns. Movement, in this sense, is not deviance. It is adaptation. It can be prudent, life-saving, and entirely rational at household level. Yet the same movement can strain the administrative systems of receiving areas, especially when arrival is sudden, categories are mixed, and local capacity was already thin. The door opened again, and the question changed from why people moved to whether institutions could absorb that movement without fraying.

This is the central tension. Mobility can reduce harm for households while increasing governance load for receiving authorities. A family that leaves a flooded district, smoke-filled town, drought-stricken farming area, or violence-prone settlement may preserve health, income, or basic safety by moving. That is the adaptive side of the story. The receiving municipality, however, may need to find beds, classrooms, language support, transport links, vaccination records, pharmacy access, sanitation, and social workers before the fiscal calendar or staffing model is ready. The concept to name here is redistributed strain. The hazard has not vanished. It has moved from one social and geographic context into another, where it is now encountered as administrative pressure.

This is why the subject should be framed through capacity rather than threat. Threat language tends to moralise movement and flatten the reasons behind it. Capacity language asks a more disciplined question: what do institutions need in order to receive people without turning reception itself into secondary harm? That question is concrete. It concerns housing stock, school flexibility, healthcare throughput, registration systems, local finance, public communication, and the ability of agencies to coordinate under uncertain numbers and changing needs. Once the focus moves there, migration appears less as a slogan and more as a planning domain. The quality of governance becomes visible in ordinary things: whether a child gets a school place, whether medicines can be refilled, whether shelters remain temporary, whether a town can expand water, waste, and transport services without political panic.

A clear example can be found in sudden displacement after wildfire or flooding, where receiving towns must absorb evacuees faster than normal market and service rhythms allow. The initial response often relies on hotels, sports halls, schools, and ad hoc shelter arrangements. That can work for days. It becomes harder over weeks or months, especially when displaced families include children who need stable schooling, older people with chronic illness, or workers whose income depends on rapid access to housing and transport. Housing pressure is usually the first visible strain. Short-term accommodation fills, rents rise, overcrowding appears, and local resentment can surface unless authorities communicate clearly and distribute burdens intelligibly. What looks from afar like a problem of numbers is often, up close, a problem of timing, coordination, and institutional flexibility.

Schooling exposes the same logic in a different register. Schools do not absorb sudden inflows simply by counting empty desks. They absorb them through transport routes, staffing, classroom capacity, language support, pastoral care, records transfer, meals, and relationships with parents under stress. A receiving district may appear to have spare room and still struggle because pupils arrive unevenly, mid-term, and with different needs. If movement follows disaster, many children carry trauma, interrupted routines, and health needs that cannot be addressed by seating plans alone. A state that treats mobility as a short technical inconvenience will often under-read the educational burden it has just shifted onto local institutions. A state that sees schooling as part of stabilisation will understand that the classroom is one of the first places where displacement becomes either governable or corrosive.

Healthcare strains reveal even more sharply why mobility is an administrative issue. Displaced households do not arrive as abstract units. They arrive with prescriptions, pregnancies, respiratory conditions, mental distress, interrupted treatments, incomplete records, and varying access to transport or money. Primary care, emergency departments, pharmacies, public health teams, and social workers all feel the pressure differently. The challenge is not merely greater demand. It is greater demand under uncertainty, with incomplete information and a time horizon nobody can define cleanly. A clinic can handle a modest rise in consultations. It is less well equipped to handle sudden demand for continuity of care when records are fragmented and housing remains unstable. Again, the issue is not whether people should have moved. Often they plainly should. The issue is whether the receiving system had enough slack and enough discretion to keep that movement from generating preventable secondary harm.

The case of internal displacement after Hurricane Maria in Puerto Rico helps illuminate this without collapsing different migration stories into one. After the storm in 2017, many Puerto Ricans moved to the mainland United States, including to Florida. The decision was often deeply practical: damaged housing, disrupted electricity, interrupted health care, schooling uncertainty, and fragile economic prospects all narrowed the viability of staying put. For households, movement could reduce immediate harm and restore access to services. For receiving communities, however, the consequences appeared in school enrolment, temporary accommodation, health provision, and local administration that had to respond before categories and needs had fully settled. The relevance here is not partisan or moralising. It is institutional. Sudden movement transformed a disaster in one place into capacity questions in another.

A second example can be found in Bangladesh, where movement from river-eroded or climate-stressed rural areas toward cities such as Dhaka has long formed part of household adaptation. The details of causation are mixed, as they almost always are: livelihoods, land loss, labour demand, kinship networks, and urban aspiration all play roles. That complexity is important. It prevents simplistic claims that climate "causes" migration in a single straight line. Yet the administrative consequence remains clear. Cities absorb people through housing markets, informal settlement, utilities, schools, clinics, and labour systems that can be stretched far beyond formal planning assumptions. Mobility may reduce risk for rural households while raising governance load in urban districts already managing poverty, congestion, flood exposure, and weak service provision. Adaptation for the household can be stress for the municipality.

That is why the chapter insists on avoiding threat framing. Threat framing does two kinds of harm at once. It can distort public perception by treating movers as the danger rather than as people managing danger. And it can degrade policy by encouraging symbolic postures where administrative design is needed. A government that responds to sudden displacement chiefly through rhetorical toughness or abstract reassurance may satisfy some audiences briefly while leaving housing, schools, clinics, and registration systems to absorb the real pressure with inadequate support.

The more useful frame is shock absorption. Which institutions absorb the movement? At what cost? With what lead time? Under which legal and budgetary constraints? How quickly can help be reallocated? Those questions do not sentimentalise mobility. They govern it.

Still, caution is necessary. Mobility should not be romanticised simply because it is adaptive. Moving can reduce one set of harms while creating others: loss of social support, debt, informal labour exploitation, school interruption, political exclusion, insecure tenure, and exposure to new hazards in low-cost land or overcrowded districts. Some households cannot move at all, which means mobility can become a stratified adaptation rather than an open one. Others move repeatedly, never reaching stability. A serious analysis therefore cannot speak as though migration is an uncomplicated solution to climate-linked stress. It is one adaptive strategy among several, and it often redistributes vulnerability rather than abolishing it. That limitation matters because it keeps the administrative focus honest. Receiving systems are not simply managing arrivals; they are managing complex forms of partial safety.

A fair counter-example also deserves notice. Not all inflows overwhelm receiving areas. Some communities and cities absorb newcomers relatively well because housing markets are less tight, local governance is flexible, civil society networks are strong, and national support arrives early enough to prevent temporary stress becoming chronic strain. Mobility can then strengthen economies, replenish labour markets, and deepen civic life. These outcomes are real. They show that migration pressure is not destiny. But they do not weaken the main argument. They sharpen it. Successful absorption is not the absence of strain. It is the presence of capacity: enough housing flexibility, school adaptability, healthcare room, and public legitimacy to convert movement from shock into manageable change.

Legitimacy matters more here than many administrative accounts admit. Residents comply with emergency housing arrangements, school integration measures, and temporary service reallocation when they believe the system is fair, finite, and competently managed. Displaced households cooperate with registration, sheltering, and transition plans when they believe those plans lead somewhere better than indefinite waiting. Public anger grows when authorities appear evasive about duration, costs, or priorities, or when locals and newcomers alike feel processed rather than governed. Trust, in this setting, is visible behaviour: whether people register, whether landlords cooperate, whether schools and clinics share information without chaos, whether host communities believe their own needs remain recognised, whether rumours outrun official communication. Administrative capacity is therefore inseparable from public credibility.

This is one reason municipal capacity becomes strategically significant. Migration pressure is often discussed at national scale, but the first durable tests occur in cities and towns. Municipal authorities manage shelters, housing enforcement, local schools, clinic referrals, water, waste, transport, and much of the everyday contact between state and household. When these institutions are underpowered, national policy begins to fail in the places where it is actually encountered. A central government may claim control while the municipality improvises beds, classrooms, permits, and sanitation arrangements with too little money and too little time. Over time that gap between national assertion and local strain becomes politically combustible. Migration then appears not as a question of movement alone, but as a referendum on whether the state can support the places doing the actual absorbing.

The phrase "migration pressure" should therefore be used carefully. Pressure does not arise simply because people move. It arises when the tempo of arrival outruns the tempo of administrative adaptation. That adaptation may involve temporary housing, payroll flexibility in schools and clinics, transport changes, emergency procurement, welfare registration, or revised local planning rules. If those instruments move quickly enough, movement may remain difficult but governable. If they do not, ordinary systems begin to jam. Queues lengthen. Informal arrangements proliferate. Political actors exploit the mismatch. A household's rational act of self-preservation is then translated, through institutional delay, into visible disorder. The danger lies not in the human movement itself, but in the failure of public systems to absorb it without humiliation or neglect.

The door opened in Yellowknife when people left danger for safety, and the same logic holds across very different settings. Mobility is one of the ways households try to stay ahead of hazards. The harder question is what happens on the other side of that door. Do receiving systems have enough room, trust, and practical authority to reduce harm further, or do they transform adaptive movement into a second emergency because housing, schooling, and healthcare were never treated as front-line infrastructure of state capacity?

Border governance and humanitarian obligation

In September 2017, as Hurricane Irma moved through the Caribbean, officials across several territories and states faced a problem that border systems are not designed to describe cleanly. The date is easy to verify. The region is clear. People moved for reasons that did not fit one legal box: danger, family networks, damaged homes, interrupted livelihoods, medical need, fear of what would happen next. The state, meanwhile, had to decide who to admit, how to register them, where to shelter them, how to screen them, and which agency owned the next step. The queue crossed the line.

That image is useful because border governance is often misrepresented as a drama of entry alone. In reality it is a sequence of administrative acts performed under uncertainty. A border is not only a line of exclusion or permission. It is a sorting system. People arrive with mixed motives, incomplete documents, uneven health needs, urgent protection claims, family obligations, and different levels of legal entitlement. Climate-linked crises intensify this complexity because they can push movement faster than categories can stabilise. A household may be escaping storm damage, income collapse, fear of insecurity, and medical interruption at once. The officer, registrar, shelter coordinator, and local government still have to act. The queue crossed the line again, and the law met ambiguity.

This is why border governance should be treated as a planning domain rather than a slogan. Public debate too often forces a false choice between moral emergency and sovereign control, as though humanitarian response and border administration occupied separate worlds. They do not. In practice they collide daily in process. Screening cannot be suspended simply because a crisis is real. Enforcement cannot function as though shelter, medical triage, and family separation risks are someone else's problem. The concept to name here is mixed-flow administration. It matters because the strategic burden lies not in choosing one principle and discarding the other, but in building institutions that can process ambiguity without collapsing into cruelty, paralysis, or theatre.

The problem begins with classification. Border systems work by category: citizen, visitor, asylum seeker, evacuee, labour migrant, irregular entrant, family member, unaccompanied child. These categories are legally significant and often operationally necessary. Climate-linked crises do not respect them neatly. A person displaced after severe flooding may also be fleeing debt, extortion, or the loss of local services. A cross-border movement after drought may involve work seeking, family reunification, and genuine fear of what staying behind now means. A government cannot simply dissolve categories because reality is mixed. But it cannot govern competently if it pretends categories fully explain reality either. The implication is severe: when legal boxes are too rigid for the conditions producing movement, the administrative system itself becomes a source of strain.

This strain is often misread as a border "security" problem in the crudest sense. Sometimes security issues do arise. The more common difficulty is throughput under pressure. Border personnel must identify vulnerabilities, verify identity where possible, refer urgent medical cases, separate those requiring immediate shelter from those who can move onward, coordinate with transport and local authorities, and do all this under legal scrutiny and public attention. These tasks are not glamorous. They are what determine whether a state appears orderly and humane or confused and punitive. When mixed flows rise quickly, the failure point is rarely the abstract concept of sovereignty. It is the desk, the shelter bed, the interpreter, the nurse, the transport manifest, the data system, the exhausted officer who has to decide which queue moves first.

A named example from Europe illustrates the problem clearly. During the arrival crisis of 2015, Greek islands such as Lesbos became sites where border enforcement, asylum processing, humanitarian triage, and local governance collided under intense pressure. The causes of movement were not climate-linked in any singular sense, and it would be analytically sloppy to say otherwise. The relevance lies in the administrative form of the challenge. Mixed flows, uneven documentation, medical needs, registration demands, shelter scarcity, intergovernmental friction, and public anxiety all arrived together. Border governance was exposed not as a single sovereign act but as a chain of procedural capacities. The case matters here because climate-linked crises can produce the same administrative collision without reproducing the same geopolitical context.

A second example comes closer to climate-linked displacement. After major storms and environmental disruption in Central America and the Caribbean, movement northward has often included people whose reasons for leaving do not fit one legal category. Storm damage, crop loss, housing insecurity, family ties, debt, violence, and the degradation of ordinary services can combine inside the same decision to move. For receiving border systems, the problem is not that these cases are mysterious. It is that they are administratively untidy. A process designed to sort cleanly between "economic" and "humanitarian" movement begins to strain when livelihoods and safety have become entangled. The officer still needs a form. The reality resists the form.

This is why process deserves more attention than rhetoric. Screening is not a mere technical prelude to the “real” policy question. It is the point at which the state decides whether it can distinguish urgency from routine without collapsing into arbitrariness. Screening under mixed-flow pressure depends on staffing, interpretation capacity, medical triage, digital systems, legal guidance, and clear escalation rules. If any of these are weak, the state may become slower, harsher, or more error-prone in ways that damage both legitimacy and actual control. A border system that cannot process complexity quickly often reaches for blunt instruments: bottlenecks, prolonged detention, ad hoc exceptions, inconsistent local improvisation, or theatrical enforcement that does little to improve throughput. The queue is then moralised instead of managed.

Sheltering reveals the same tension in a different form. Border shelters are often imagined as temporary humanitarian spaces, but in practice they are political and administrative filters. They determine whether new arrivals remain visible to the state, whether health needs are stabilised, whether families can rest and receive information, whether trafficking or exploitation risks increase, and whether onward movement becomes more orderly or more desperate. A sheltering system under pressure can very quickly become a legitimacy test. If facilities are overcrowded, unsafe, medically thin, or administratively disconnected from registration and onward planning, the public sees disorder and the displaced experience the state as a holding pattern rather than a governing presence. Humanitarian obligation then looks, to critics, like institutional chaos, when the deeper problem is underbuilt process.

Coordination across levels of government is where these stresses often become strategically significant. National ministries may control border rules, but municipalities handle shelter overflow, local policing, schooling, clinics, sanitation, and much of the public reaction. Regional authorities may manage transport hubs or health systems. Civil society and international organisations may provide staff, food, or medical support while lacking formal authority. If roles are unclear, costs are shifted downward, or data do not move between levels, the border becomes a national symbol resting on local improvisation. That is a fragile arrangement. A central government may sound firm while a municipal authority struggles to house families in gymnasiums and keep local services from buckling. Sovereignty performed at the line can conceal administrative weakness behind it.

This is one reason humanitarian obligation cannot be treated as a moral add-on to border policy. It is part of operational credibility. A person who is dehydrated, ill, traumatised, separated from family, or unable to understand instructions is harder to process lawfully and safely if shelter and triage are inadequate. Children, pregnant women, older adults, and those with chronic illness do not wait for legal categories to stabilise before requiring care. A state that neglects this under the banner of control often creates more disorder, not less, because untreated vulnerability slows processing, raises error rates, invites litigation, and fuels public scandal. Humane process is not the opposite of administrative order. It is one of its conditions.

Still, caution is necessary. Not every border pressure should be reframed as a climate issue, and not every ambiguous flow is best handled through expansive humanitarian logic. States have legitimate interests in identification, screening, security checks, and the enforcement of immigration law. The argument here does not erase those interests. It insists that enforcement systems built around narrow, binary assumptions are poorly suited to climate-linked crises in which mobility may be sudden, mixed in motive, and administratively messy. A serious analysis avoids two opposite mistakes: treating all movement as innocent adaptation beyond legal process, or treating all mixed movement as strategic threat demanding performative hardness. Both weaken governance by substituting posture for design.

A useful counter-example lies in systems that have built stronger reception-and-processing linkages without abandoning legal screening. Some jurisdictions have shown that early medical triage, clearer registration pathways, temporary protected statuses in specific circumstances, and better national-local coordination can reduce institutional chaos while preserving state authority. The specific legal instruments differ, and no single model fits every constitutional order. The point is simpler. Border systems function better when they accept that mixed flows require procedural elasticity rather than rhetorical escalation. Elasticity here does not mean softness. It means the capacity to sort, shelter, and decide without pretending that a single queue contains only one type of human circumstance.

Trust matters sharply in this domain because border governance operates under intense public scrutiny and often deep political suspicion. Residents near reception areas need to believe that local services will not simply be abandoned to absorb costs alone. Border staff need to believe that difficult discretionary judgments will be backed by coherent guidance rather than punished opportunistically after the fact. New arrivals need to believe that registration leads somewhere other than indefinite uncertainty. If these forms of trust weaken, cooperation frays. People avoid registration, local authorities resist national directives, and front-line staff default to defensive behaviour. What then appears as loss of control may in fact be the result of a state that has not invested enough in making control administratively believable.

This is especially important during climate-linked crises because time horizons collapse. A storm, crop failure, or cascading service breakdown can move people before legislatures have revised categories, before municipalities have secured funding, before ministries have agreed who pays for shelter, and before public narratives have adjusted. Border systems are then judged by the quality of their first administrative moves. If those moves are slow, contradictory, or visibly inhumane, later correction may come too late to prevent political damage. Process matters because it is the state's first language under pressure. Long before strategy papers catch up, the screen, the shelter, and the coordination call are already telling the public what kind of system exists.

The temptation to overstate direct links between climate and border crisis should also be resisted here, because exaggerated claims can degrade policy design. If every mixed flow is presented as climate migration in pure form, legal categories become less credible and evidence discipline weakens. If climate is ignored entirely, planners miss how environmental disruption can intensify ambiguity by destabilising livelihoods and safety at the same time. The disciplined position is narrower and more useful. Climate-linked crises can make border administration harder not by abolishing law, but by producing people whose reasons for moving are entangled in ways that test systems built for clearer distinctions. Governance must begin from that administrative reality.

The queue crossed the line in the Caribbean after Irma, on Greek islands in 2015, and in many quieter places where legal categories met urgent human movement under strained conditions. That is why migration should be treated here not as a slogan about openness or closure, but as a planning problem of throughput, triage, shelter, and coordination. The harder question is not whether the state can enforce a line in principle. It is whether, when the next mixed flow arrives under climate-linked stress, the line will be backed by institutions capable of processing human ambiguity without turning procedural weakness into moral spectacle or political drift.

Urban absorption and informal settlement risk

In August 2005, Hurricane Katrina struck the Gulf Coast of the United States and displaced large numbers of people, many of whom moved into other cities and towns already carrying their own housing and service burdens. The place was New Orleans at the point of shock, but the administrative story travelled far beyond it. The date is easy to verify. The immediate drama belonged to flood, evacuation, and visible devastation. The quieter and more durable question was urban: what happens when cities become the place where another region's crisis must be absorbed before budgets, housing markets, and public services are ready? The edge fills first.

That image matters because cities are often treated as if they possess an almost natural capacity to absorb human movement. They have jobs, informal economies, transport links, clinics, schools, landlords, markets, and social networks. People arrive because cities offer options when rural systems fail, coasts become dangerous, conflict worsens, or livelihoods break down. In that sense, cities do act as shock absorbers. They can spread risk across larger labour markets and service systems. They can reduce household harm by offering anonymity, income chances, and proximity to care. But a shock absorber can also wear out. When inflows are rapid, prolonged, or layered onto pre-existing scarcity, the city's absorptive role begins to backfire. The edge fills first again, this time not on a map but in the budget, the waiting room, the school gate, and the drainage ditch beside a new informal dwelling.

This is the central tension. Urban absorption can protect people while exposing infrastructure and politics to new strain. A household arriving in a city may be safer than it was in a drought-hit district, eroding village, smoke-affected town, or violence-prone region. The city may lower immediate risk by providing shelter of some kind, access to markets, and better odds of medical attention or schooling. Yet that same arrival can increase governance load on municipal systems already stretched by housing shortages, weak utilities, informal transport, and thin administrative staffing. The concept to name here is municipal shock absorption. The implication is exacting: the city becomes a frontline institution for national stability, even when national governments still speak as if migration were mainly a border or humanitarian issue.

Housing is usually the first and clearest pressure point. New arrivals do not wait for zoning plans, capital budgets, or serviced land releases. They go where cost, kinship, transport, and immediate possibility allow. Formal rental markets may absorb some. Many cannot. Informal subdivisions, overcrowded rooms, improvised peri-urban settlements, and structures built on marginal land become the practical answer. This is not merely a housing story. It is the beginning of a new risk geography. When settlement expands onto floodplains, unstable slopes, coastal fringes, drainage corridors, or polluted land, the city solves one problem by planting the seeds of another. The urban fringe becomes a ledger written in corrugated metal, plastic piping, informal electricity, and future emergency spending.

Dhaka offers a powerful example, not because it stands alone, but because it makes the logic hard to miss. The city has long absorbed people moving for many reasons: river erosion, rural poverty, labour demand, social networks, and environmental stress among them. The result has not been one simple narrative of climate migration. It has been a sustained test of urban capacity. Informal settlement and low-cost housing have spread into areas where drainage, sanitation, transport, and flood management are already difficult. The relevant point here is not to reduce Dhaka to vulnerability alone. It is to see how municipal governance becomes nationally strategic when the city functions as the default absorber of displacement and livelihood stress. National policy may speak of development, adaptation, or migration management. The practical test appears in drains, clinics, schools, and tenancy enforcement.

Public health risk follows quickly once urban growth outruns service provision. A city can absorb additional population for a time through ingenuity, informal exchange, and simple crowding. It becomes far harder to absorb when water supply, waste disposal, sanitation, and primary care begin to lag. Overcrowded housing complicates disease control. Informal water access changes exposure to contamination. Drainage failure turns seasonal rain into recurrent local emergency. Public health systems then face a double burden: more demand from more people, and more complex conditions because the built environment is growing in ways the formal city does not fully govern. What later appears as sudden outbreak, dangerous flooding, or neighbourhood unrest is often the visible form of a slower administrative mismatch.

This is why municipal capacity should be understood as a strategic variable rather than a local technical concern. A city that cannot process housing pressure, school enrolment, clinic demand, waste management, and informal land occupation under sustained inflow does not merely face a planning difficulty. It becomes a national political risk. Residents see crowding, slower services, rising rents, and deteriorating trust in public promises. Newcomers see insecurity, exclusion, and often a state present mainly through eviction threats or policing rather than through service. Political actors then find fertile ground for polarisation. The city ceases to be a quiet absorber and becomes a visible theatre in which anxieties about fairness, belonging, and state competence are rehearsed daily. The edge fills first in public imagination as well.

A second example can be found in Karachi, where migration, displacement, labour demand, and political fragmentation have long interacted with informal urban expansion. As with Dhaka, the causes of movement are mixed and should not be flattened into a single climate story. That is precisely the point. Municipal strain rarely arrives with one pure cause attached. It comes through layered pressures that accumulate in land markets, settlement patterns, transport systems, and unequal service provision. Karachi demonstrates how informal settlement can become both an adaptive solution for households and a generator of deeper exposure when housing spreads into land poorly served by drainage, sanitation, or emergency access. The city absorbs and destabilises in the same movement.

This is one reason the language of "informality" requires caution. It can sound like a neutral description or a moral judgement depending on who uses it. Here it should be read as a governance condition. Informal settlement often emerges because formal systems are too slow, too expensive, too exclusionary, or too underbuilt to absorb real demand. It is therefore not simply a failure of compliance. It is a signal that the city is governing movement through tolerated improvisation rather than planned provision. That improvisation can be life-saving and economically productive. It can also shift risk downward and outward, concentrating new residents in places where flood exposure, fire risk, heat stress, or service deficits are higher. To condemn informal settlement without asking what formal system failed to provide an alternative is analytically lazy. To celebrate it without accounting for the risks it can intensify is equally shallow.

The brief example of flood exposure increasing when housing spreads into marginal land captures this dynamic neatly because it shows how adaptation at one scale can manufacture vulnerability at another. A family settling on low-cost flood-prone ground at the edge of a city may be acting rationally. The location offers access to work, transport, or kin and is affordable in a way safer land is not. The municipal system, however, may inherit a different set of consequences: blocked drainage, repeated emergency rescue, disease risk after inundation, school closures, road damage, and future political conflict over upgrading versus relocation. The household solves immediate survival. The city acquires a recurring hazard and a future legitimacy test. This is not a moral accusation. It is how urban shock absorption works when land, law, and housing affordability are badly misaligned.

Rapid growth also strains political patience. Residents already in a city do not experience absorption as an abstract demographic process. They experience higher rents, longer queues, more crowded buses, slower clinics, and schools under pressure. If governments cannot explain how they are managing the inflow, or if they appear to neglect long-term residents while failing to protect newcomers adequately either, resentment can sharpen into polarisation. The danger here is not only prejudice, though prejudice matters. It is institutional mistrust. Once public services begin to look incapable of absorbing demand fairly, political actors can turn administrative strain into stories of invasion, betrayal, or elite indifference. The city's infrastructural stress is then converted into national grievance.

Still, caution is necessary. Rapid urban growth does not inevitably produce collapse or extremism, and not every informal settlement is a prelude to crisis. Some cities absorb newcomers better than outside observers expect because they possess flexible economies, social networks, local ingenuity, and administrative arrangements that, while imperfect, can adapt faster than formal models assume. Informal districts can become stable neighbourhoods over time. Incremental upgrading can work. Schools and clinics can stretch intelligently. This matters because the argument here is not that urban migration pressure is inherently destabilising. It is that municipal capacity determines whether absorption remains adaptive or begins to feed back into new hazards and political strain.

That limitation sharpens rather than weakens the strategic point. If outcomes vary, then capacity is doing explanatory work. One city integrates growth with enough housing flexibility, local finance, transport adaptation, and service extension to remain governable. Another sees the same broad pressures translate into dangerous land occupation, water stress, repeated flooding, schooling deficits, and polarised politics. The difference is not simply wealth. It lies in land governance, local fiscal room, administrative competence, infrastructure planning, and the degree to which national governments treat municipalities as real front-line institutions rather than as places where national problems are deposited. Municipal capacity, in that sense, is not a local detail. It is a national security asset wearing local clothes.

Health systems reveal this with special force. A city can tolerate informal housing growth for some time while still appearing stable. The point at which stability begins to fray is often seen in public health: diarrhoeal disease after drainage failure, respiratory illness in overcrowded settlements, heat risk in poorly serviced housing, maternal care disruption when clinics cannot keep pace, longer emergency response times where roads are narrow or unplanned. These are not only medical issues. They are indicators of whether urban absorption remains within the bounds of governable adaptation. Once health impacts become visible and unevenly distributed, trust in city government and in the fairness of growth begins to change. The edge fills first in the clinic too.

A useful counter-example lies in places where municipal authorities, despite fiscal limits, have invested in serviced land, incremental upgrading, local health outreach, and transport or drainage extensions before settlement hardens into chronic neglect. Such cases are rarely perfect and often remain vulnerable to larger economic or climatic pressures. But they show that cities need not choose only between exclusion and unmanaged sprawl. The relevance here is not to prescribe one urban model. It is to insist that when municipalities are given room, data, legal tools, and money early enough, they can turn inflow into planning rather than emergency improvisation. Where they are denied those tools, the burden does not disappear. It shifts into informal risk.

This is why the chapter treats cities as shock absorbers that can backfire. A good shock absorber takes impact and preserves control. A failing one transmits the force elsewhere in distorted form. Urban systems do something similar. They can absorb movement and distribute it through labour markets, neighbourhoods, and services in ways that lower social damage. Or they can transmit the shock into flood-prone land, overwhelmed clinics, unsafe tenancies, school strain, and a politics increasingly organised around visible scarcity. The same city can do both at different moments or in different districts. What matters is not simply the arrival of people. It is whether administrative systems are fast and trusted enough to keep the edge from becoming the city's default planning instrument.

The edge filled first in Dhaka and Karachi and in countless less famous cities where migration, hazard, and uneven governance are writing the future of national stability in municipal form. The harder question is not whether cities will continue to absorb people under climate-linked stress. They will. The question is whether states will treat municipal housing, drainage, clinics, schools, and land governance as strategic infrastructure worthy of national seriousness - or whether they will continue to discover, one floodplain and one overcrowded district at a time, that the place where the city took people in is also the place where the next crisis has already begun.

Conflict risk claims and evidence discipline

In 2011, as protest spread across Syria and then hardened into civil war, a familiar claim quickly gained international traction: drought had led to migration, migration had led to unrest, and climate had therefore helped cause conflict. The date is easy to verify. The place is unmistakable. The appeal of the claim was obvious. It offered a clean story, a modern warning, and a direct route from environmental stress to strategic breakdown. It was also too neat. The line was drawn too straight.

That straight line remains one of the central temptations in climate-security argument. A severe drought, a failed harvest, a wave of displacement, a city under strain, a protest movement, a crackdown, armed mobilisation: place them in sequence and the result appears legible, almost cinematic. Yet serious analysis has to ask harder questions. Which of these links are evidenced and which are assumed? Which pressures altered livelihoods, which reshaped grievances, which weakened service provision, which sharpened elite competition, and which provided openings for actors already willing to use coercion or violence? Once those questions are asked properly, the climate-conflict story often becomes more complex and more useful at the same time. The line was drawn too straight because institutions wanted simplicity more than accuracy.

This matters because over-claiming direct links between climate and conflict does not merely offend scholarly caution. It distorts policy. If climate is treated as a direct cause of violence in cases where the real pathways run through governance failure, uneven service provision, elite manipulation, corruption, policing choices, or opportunistic armed actors, then public action can drift towards the wrong remedies. One ends up speaking as though rainfall, temperature, or mobility were themselves the political adversary, when the decisive processes lie in how states and societies manage strain. The concept to name here is pathway discipline. It asks analysts to trace how environmental pressure may alter the conditions under which conflict becomes more plausible, without pretending that physical stress acts with political intention of its own.

The disciplined approach begins with grievance. Grievances are not epiphenomenal emotions drifting above material life. They are structured perceptions of exclusion, unfairness, humiliation, or abandonment, and they matter when they are attached to visible institutional behaviour. A harvest failure may hurt many people. It becomes politically dangerous in a particular way when compensation is selective, relief is captured by patronage networks, land rights are ignored, or local authorities appear absent or predatory. Climate-linked stress can intensify grievances by making scarcity sharper or recovery slower. It does not automatically generate collective violence. The implication is precise: one must look not only at hardship, but at who is seen to absorb it, who is seen to escape it, and what administrative choices translate suffering into resentment.

Service failure is the next pathway, and often the most underappreciated. Political order depends less on abstract legitimacy than on everyday proof that the state can still keep water running, roads open, clinics functioning, schools staffed, and prices or supplies from spiralling completely out of reach. Environmental stress can weaken these services directly or indirectly. Drought presses water systems and rural livelihoods. Flooding damages roads, electricity, and clinics. Heat strains health services and power grids. Displacement pressures housing, sanitation, and schooling. None of this produces conflict by itself. Yet once services begin to fail unevenly or repeatedly, public trust can narrow and local dispute can become easier to mobilise. The line to violence is still not automatic. But the administrative terrain becomes more permissive for those who want to exploit it.

Elite competition is where that permissive terrain often becomes decisive. Political and economic elites do not passively receive climate stress. They interpret it, instrumentalise it, and sometimes intensify it. A shortage can become a patronage opportunity. A displaced population can become a voting bloc, a policing target, or a scapegoat. Relief funds can be channelled strategically. Land use decisions can favour allies. Security forces can respond selectively. Opposition movements can frame service collapse as proof of regime failure. In such contexts, the climate variable matters because it has increased pressure on the system, not because it has ordained the political outcome. The serious claim is therefore conditional: environmental stress can sharpen the stakes of elite competition by making distributional conflict more consequential and failures of governance more visible.

Opportunistic violence completes the pathway. Armed groups, criminal networks, militias, and predatory political entrepreneurs often thrive not where climate has "caused conflict", but where stress has weakened oversight, stretched security forces, disrupted livelihoods, and increased the value of coercive control over movement, aid, transport, or land. A floodplain settlement may become vulnerable not because flood makes people violent, but because weakened policing and land informality create openings for extortion or political intimidation. A drought-hit area may see more recruitment not because thirst becomes ideology, but because income loss and state neglect lower the cost of joining an armed actor offering money, protection, or belonging. Violence, in this reading, is not the direct child of climate. It is the work of people using stress conditions to pursue advantage.

The Syrian case is useful precisely because it demonstrates both the value and the limits of climate-conflict claims. Environmental stress mattered. Drought affected livelihoods and mobility in ways that formed part of the wider social picture. But the path from environmental pressure to civil war ran through state policy, rural neglect, migration handling, political repression, elite choices, and escalating violence by organised actors. To say that climate "caused" the war is analytically careless. To say that environmental stress contributed to the conditions within which grievances, displacement, and state weakness became more combustible is far more defensible. The distinction is not semantic pedantry. It changes what one thinks should have been monitored, supported, or reformed before violence escalated.

A second example can be found in farmer-herder conflicts discussed in parts of the Sahel and Nigeria. Public debate often compresses these cases into a direct story of desertification or climate change producing communal violence. Environmental pressures do matter in many such settings, including through changing access to land, water, and pasture. But again, the conflict pathway is more crowded than the slogan allows. Land tenure systems, arms availability, policing weakness, corruption, local histories of grievance, political brokerage, and opportunistic narratives all shape whether resource stress becomes violent confrontation, tense coexistence, or negotiated adaptation. The disciplined analyst therefore resists both denial and determinism. Stress can raise pressure. Institutions and actors decide how much of that pressure becomes violence.

This is why evidence discipline matters so much. Climate-security argument is unusually vulnerable to rhetorical inflation because the stakes are high and the topic travels easily across policy audiences. A straight-line claim from climate to conflict sounds urgent, modern, and morally serious. It is also often hard to test cleanly, because the cases most often cited are already marked by many overlapping forms of fragility. Evidence discipline asks different questions. What changed materially? What changed institutionally? What forms of grievance became more salient? Which services failed or were perceived to fail? Which elite actors benefited from escalation? Which violent organisations exploited the opening? Which parts of the story are evidenced directly, and which remain plausible but uncertain? This approach may produce messier narratives. It also produces more reliable ones.

A common objection is that such caution risks understating danger. If analysts become too fastidious, do they not miss the political significance of worsening environmental conditions? The concern is understandable. Yet the greater danger usually lies in the opposite direction. Over-claiming can damage credibility, especially when later evidence shows the direct link was weaker, more contested, or more politically mediated than first suggested. It can also encourage broad, symbolic policy responses while neglecting the very institutional weak points through which environmental stress actually becomes dangerous. If the real pathway runs through municipal service collapse, land administration, policing bias, or the fiscal exhaustion of receiving cities, then shouting "climate conflict" may generate attention while blurring the site of intervention.

This does not mean climate should be pushed to the margins of security analysis. It means it should be located correctly. Climate-linked stress belongs in conflict analysis when it alters livelihoods, movement, scarcity, service reliability, state credibility, or the stakes of elite competition in ways that plausibly affect violence risk. It does not belong there as a universal causal solvent. Pathway discipline allows one to say something both serious and careful: that environmental pressures can help create conditions in which conflict becomes more likely, but only through social, political, and institutional mechanisms that must be traced rather than presumed. This is more demanding than the simple story. It is also far more useful to those charged with prevention.

The discipline can be made practical. When evaluating a claim that climate stress is raising conflict risk, one should ask which grievances are becoming more salient and for whom; whether public services are degrading and where; whether displaced or newly mobile populations are interacting with strained urban or border systems; whether elite competition is sharpening around relief, land, or security; and whether armed or coercive actors are exploiting the resulting openings. If these pathways are weak or absent, the climate-conflict claim should remain modest. If several are visible and interacting, the claim can be stronger without becoming simplistic. The method guards against both sensationalism and complacency.

A fair counter-example should be kept in view. Some places endure severe climate-linked stress without descending into violence because institutions, however imperfect, preserve channels of mediation, service continuity, and political bargaining. Communities adapt. Elites compromise. Security forces remain restrained enough. Local grievances are real but do not harden into organised violent mobilisation. These cases matter because they show that hardship alone does not dictate outcome. They also reveal where capability resides. Conflict prevention under environmental stress often depends less on predicting climate perfectly than on preserving administrative fairness, credible service, and space for negotiation before coercive actors define the situation first.

This is also why migration must not be smuggled into conflict analysis as a stand-in for danger. Movement can alter local politics and strain urban systems, as the previous section argued. It can also distribute risk, sustain households, and reduce immediate harm. Whether migration contributes to violence depends on how it intersects with housing, labour, municipal services, elite narratives, policing, and informal settlement governance. To treat movement itself as a conflict signal is to repeat the very analytical laziness this section is trying to correct. What matters is not the presence of mobility alone, but the institutional conditions through which it is received, framed, and governed.

The deeper strategic gain from evidence discipline is that it preserves trust in security analysis itself. Policymakers, military planners, humanitarian agencies, and municipal officials all need language that is strong enough to warn and careful enough to guide. If climate is made to explain too much, it will soon explain too little. If every case of unrest in a stressed environment is labelled climate conflict, the term becomes rhetorically potent and analytically weak. By contrast, a pathway-based account can tell us where the real pressure points lie: the failing clinic, the underfunded municipality, the politicised relief channel, the land registry that cannot cope, the border shelter that has become a bottleneck, the local leader who turns scarcity into grievance and grievance into mobilisation.

The line should not be drawn too straight. That is the real lesson. Climate-linked stress can matter profoundly in conflict settings, but it matters through people, institutions, and choices. The unresolved question is therefore not whether climate causes conflict in some abstract universal sense. It is whether analysts and governments can discipline themselves to identify the pathways accurately enough that the next warning does not arrive wrapped in a story too simple to support the policy it demands.

Chapter 9

Risk Communication and Strategic Trust

What makes a warning believable when the danger is real but the shape of it is still changing? That question lies closer to state capacity than many governments care to admit. Public institutions often speak as though information were a neutral good: gather more of it, publish enough of it, and trust will follow. Experience suggests otherwise. A warning can be accurate and still fail. Guidance can be cautious and still sound evasive. A revision can be responsible and still feel like a confession of incompetence. The microphone is live.

That is why risk communication belongs inside the architecture of governance rather than at its polished outer edge. Institutions do not merely describe danger; they help determine how danger is socially organised. They signal who is expected to act, what kind of uncertainty remains, what level of inconvenience is justified, how long sacrifice may be required, and whether later changes in guidance should be understood as correction or betrayal. If that signalling is inconsistent, late, overconfident, or thinly explained, even technically sound policy can begin to fail in public. The microphone is live again, and this time the issue is not just what is said but whether the speaker still sounds like a credible steward of uncertainty.

This creates a paradox at the centre of modern public authority. Institutions are asked to project confidence and competence precisely when the evidence is incomplete, the hazard evolving, and the operational picture still unstable. Speak too firmly and later revision will look deceptive. Hedge too much and urgency dissolves into fog. Say too little and rumour fills the gap. Say too much without role clarity and different agencies may contradict one another in public. Strategic trust therefore cannot mean the elimination of uncertainty from the message. It means building a voice that can survive revision without sounding fraudulent and can issue clear instruction without pretending to know what is not yet knowable.

That tension matters because trust is not a moral ornament added after the real work is done. It is one of the conditions under which the real work succeeds. Evacuations proceed more smoothly when people believe routes and shelters are being described honestly. Public health guidance holds when updates are expected rather than treated as reversals of principle. Infrastructure warnings are taken seriously when previous warnings were not theatrical. Staff inside government are more likely to speak candidly when they trust that political leaders will not punish every visible adjustment as weakness. Strategic trust, in this sense, is institutional room to move under pressure without losing the audience needed to make movement matter.

The stakes extend further once climate-related issues become identity markers rather than shared administrative problems. Expertise can then be pulled into a culture war not because the underlying evidence changed, but because its public meaning did. Analysts are harassed, local officials become more cautious, routine technical judgments are recoded as partisan signalling, and policy begins to lurch between overstatement and retreat depending on who last held the microphone. The damage is not only rhetorical. Delayed action, staff intimidation, and policy whiplash all impose operational costs. When institutions cannot protect the conditions of analytic integrity, communication ceases to guide action and begins to reproduce faction.

Modern media environments intensify this exposure because they alter tempo. Panic, complacency, scapegoating, false reassurance, and selective outrage can now spread with a speed that administrative systems rarely match. That does not mean the public is irrational or that all misinformation is external malice. It means that communication has become part of emergency management itself. A rumour about a closing road, a false claim about shelter conditions, a distorted clip of an official update, or a confident but wrong interpretation from a partisan intermediary can change behaviour before formal correction arrives. Institutions that still treat communication as the final stage of policy, rather than one of the operational conditions of policy, will find themselves perpetually late to the reality they are trying to govern.

The path ahead follows that problem from voice to environment and then to legitimacy. It begins with the difficulty of speaking credibly about uncertain but severe risks, moves into the political pressures that warp expertise and make consistency harder to sustain, then turns to media systems as accelerants that can enlarge fear or denial faster than official channels can respond. From there it settles on the slower work by which trust is built before crisis through participation, consultation, and the visible integration of public constraints into planning, without pretending that consultation is costless or automatically wise.

What joins these themes is the problem of role clarity. Citizens do not need every institution to say everything. They need institutions to know who explains uncertainty, who gives operational instruction, who updates evolving guidance, who speaks for the evidence base, and who is accountable when those functions collide. Confusion about role is one of the quickest ways to burn trust. So is the opposite error: centralising the voice so completely that local knowledge, municipal reality, and frontline constraints disappear from the message. Strategic communication requires coordination without ventriloquism. It needs a state voice that is coherent enough to command confidence and plural enough to remain connected to conditions on the ground.

This is also why legitimacy cannot be treated as something presumed into existence by office. It is produced through repeated encounters in which institutions show that they can listen, revise, explain, and still decide. Consultation can reveal hidden constraints, expose trade-offs early, and widen compliance later because people recognise their own realities in the plan. It can also consume time, invite performative participation, and create frustration when the scope of change is unclear. The danger lies in both extremes: assuming legitimacy arrives automatically with authority, or imagining it can be manufactured endlessly through process alone. What matters is whether participation changes the quality of the plan enough that, when disruption comes, people experience the state not as a distant broadcaster but as a system that already knows something about the lives it is asking to bend.

The microphone remains the image worth keeping. It amplifies clarity, but it also magnifies confusion, inconsistency, vanity, and fear. Once a crisis begins, every delayed update, every contradiction, every unexplained change in tone becomes part of the hazard environment itself. The harder question, then, is not whether governments can communicate more. It is whether they can build a voice credible enough to tell the truth under uncertainty, durable enough to survive revision, and trusted enough that when the next instruction arrives through the static, people still hear not just information, but authority worth following.

Uncertainty, credibility, and institutional voice

In August 2023, authorities in the Northwest Territories ordered the evacuation of Yellowknife as wildfire threatened the city. The place was northern Canada. The date is easy to verify. The danger was real, but the picture was not stable. Fire behaviour, smoke, road conditions, aircraft availability, and the likely pace of change all shifted as officials tried to speak clearly enough to move people without sounding either panicked or falsely certain. The message had to move.

That image is useful because institutions do not fail in risk communication only when they lie. They also fail when they speak as though uncertainty were either shameful or politically unbearable. A severe but evolving threat creates a peculiar burden. Officials are expected to sound calm, informed, and decisive, even when the evidence is incomplete and operational conditions remain fluid. Speak too firmly and later revision can look like deception. Hedge too much and the warning may dissolve into administrative mist. Delay the message while awaiting greater clarity and the public may infer either incompetence or concealment. The message had to move again, and with it the public's judgment about whether the state still knew what it was doing.

This is the central tension. Institutions struggle to speak about uncertain but severe risks because public authority is still culturally tied to a model of confidence that uncertainty keeps disrupting. Government is expected to know, or at least to sound as though it knows. Yet climate-linked emergencies, compounding shocks, infrastructure stress, and evolving public health conditions often do not present themselves in final form at the moment action is needed. The concept to name here is authoritative uncertainty. It refers to the problem of exercising legitimate direction while openly acknowledging that the hazard, the evidence, or the likely consequence is still changing. The implication is exacting: credibility no longer depends on projecting certainty at all costs. It depends on how institutions structure, update, and limit their claims without surrendering operational clarity.

This is why overconfidence is so dangerous. Overconfidence does not merely raise the risk of later embarrassment. It changes the meaning of revision when revision becomes necessary. If a minister, mayor, public health leader, or emergency official speaks as though a forecast were settled, a route secure, a shelter adequate, or a disruption temporary, and the next update contradicts that assurance, the public does not only learn that conditions changed. It learns that the institution asked to be trusted in one tone and returned in another. The damage is not confined to the sentence that proved too strong. It spreads across later instructions. A revised evacuation boundary, a second shelter directive, or a new service warning is then heard through the memory of the earlier certainty that did not hold.

Hedging can be just as corrosive, though in a different way. An institution that fears being wrong may speak in terms so cautious that action becomes hard to infer. "We are monitoring closely." "The situation remains dynamic." "All options remain under consideration." Such phrases are not false. They are often administratively empty. They communicate awareness without communicating consequence. In an evolving emergency, this can be as damaging as false firmness because it asks the public to tolerate uncertainty without telling people what uncertainty should change in their behaviour. A warning that leaves action unclear shifts the burden of interpretation onto individuals, rumours, and private judgement. That is often where trust begins to fragment.

The example of changing guidance during a rapidly developing emergency reveals why consistent framing matters so much. Suppose officials initially say that residents in one area should prepare to leave but need not evacuate yet. Hours later, the instruction changes: leave now. The revision itself is not the problem. The problem arises if the first message implied stable safety rather than conditional assessment. A public that was told, in effect, "you are safe" may hear the second instruction as contradiction. A public that was told "you are not under order yet, but the trigger for leaving may be reached quickly if conditions worsen" is more likely to hear the revision as confirmation of a process already explained. The difference lies not in data alone but in framing. Updating guidance without appearing deceptive requires institutions to tell people at the outset what kind of change in evidence or conditions would alter the instruction later.

This is where clear roles become indispensable. Public communication often falters because institutions try to make one voice do several incompatible jobs. A technical agency wants to describe uncertainty precisely. A political leader wants to maintain calm and visible control. An operational commander wants the public to comply with a simple instruction. A local authority wants nuance because its conditions differ from the regional map. None of these roles is illegitimate. Confusion begins when they blur. If scientific explanation sounds like command, if political reassurance contradicts operational caution, or if local detail is swallowed by a central slogan, credibility weakens even when each individual speaker believes they are helping. A trustworthy institutional voice is therefore rarely a single voice. It is a coordinated arrangement in which different roles remain distinct enough that the public can tell who is explaining evidence, who is issuing instruction, and who is accountable for revision.

A named example from outside climate but highly relevant here is the changing public guidance issued during the COVID-19 pandemic. The topic became politically toxic in many places, yet its communication lesson remains clear. Guidance that changed was not, by itself, evidence of incompetence. Under evolving evidence and changing conditions, revision was inevitable. The damage arose when institutions presented early judgements as settled fact, failed to explain why later revisions followed from new information, or allowed several authorities to speak in tones so inconsistent that the public began to hear all updating as either excuse or manipulation. The relevance to climate-linked risk communication is exact. The issue is not whether institutions should revise. They must. The issue is whether revision has been built into the public voice before the crisis makes it visible.

A second example can be found in wildfire emergencies, where smoke, wind shifts, route viability, and suppression outcomes change quickly. Here, the public does not need the certainty that no official can provide. It needs a stable communication architecture. What is the current instruction? What conditions would change it? Who should move first? Which routes are preferred now, and which conditions would make them unusable? Where will the next verified update appear? What is not known yet? If these elements remain consistent even while the details shift, credibility can survive repeated updates. If the tone lurches between assurance and alarm, or if agencies contradict one another on the basic question of who is speaking for what, the same hazard becomes harder to manage because the communication system has become part of the emergency rather than a support to it.

This is why credibility should be understood less as personal charisma than as institutional consistency over time. Citizens do not decide whether to believe a warning purely by inspecting the moral fibre of the spokesperson. They judge patterns. Did earlier warnings prove useful? Were later corrections acknowledged openly? Did agencies keep to their roles? Were thresholds explained in advance? Did changing guidance correspond to observable changes, or did it appear politically driven? Were local conditions incorporated without dissolving the coherence of the wider message? Credibility accumulates through these repeated encounters. It is a public memory of whether the institution's voice tends to survive contact with changing reality.

Still, caution is necessary. No communication design can eliminate distrust completely, and no institution can speak in a way that satisfies every audience under severe uncertainty. Some people will always read revision as weakness or deceit. Some hazards evolve too quickly for full explanation before action is needed. Some operational details cannot be made public without undermining the response itself. A serious argument should admit these limits. The claim is not that careful framing guarantees trust. It is that bad framing burns trust much faster than uncertainty alone does. Overconfidence and vagueness are not inevitable properties of crisis communication. They are habits, and habits can be changed.

This matters strategically because communication is not merely the final layer placed over policy once decisions are made. It is one of the conditions under which decisions become effective. Evacuation depends on what people hear and when they hear it. Fuel conservation, sheltering, school closure, infrastructure warnings, and hospital pressure all become publicly governable only through messages that connect uncertainty to action without either masking the uncertainty or surrendering the action. An institution that cannot manage its own voice under stress often discovers that its operational capacity is weaker than its equipment and plans suggested. The message had to move, and the system moved with it.

The phrase "transparent updating" is useful here because it names the practice most likely to preserve credibility without pretending away instability. Transparency does not mean flooding the public with raw internal debate or technical ambiguity. It means stating what has changed, what has not changed, why guidance is being revised, which role is speaking, and when the next update can be expected. It means treating revision not as a public-relations accident but as part of the official process. A government that can do this teaches the public how to interpret changing guidance before the next shock arrives. A government that cannot will keep discovering, too late, that each revision has been heard as a confession rather than as a competent response to new conditions.

This is why institutional voice is a governance asset. It is built long before the crisis briefing. It is built in everyday consistency, in the absence of theatrical certainty, in the habit of naming uncertainty without surrendering direction, and in the discipline of letting different offices perform different communication roles without collapsing into cacophony. The hard question is not whether institutions can be trusted in the abstract. It is whether, when the next severe but evolving risk emerges, they have built a voice capable of saying "this is what we know, this is what may change, this is what you should do now" in a way that sounds neither evasive nor omniscient - and therefore still worth following when the next update inevitably comes.

Polarisation and the politics of expertise

In August 2019, when Brazil's National Institute for Space Research became the focus of political dispute over deforestation data, the episode exposed a problem larger than one country and one institution. The organisation was INPE. The place was Brazil. The date is easy to verify. What mattered was not only disagreement over the figures themselves, but the speed with which environmental measurement became a test of political identity, loyalty, and narrative control. Once that happened, the argument stopped being only about evidence. It became a struggle over who was authorised to describe reality at all. The microphone changed hands.

That image is useful because climate issues rarely remain confined to technical assessment. They acquire symbolic charge. A warning about fire risk, a floodplain map, a drought projection, a heat-health advisory, a relocation study, or a resilience budget can all become markers in wider identity contests. People do not then receive the information as a claim to be weighed alongside others. They receive it as a sign of who is speaking, what tribe they belong to, and which social or political order the information seems to affirm or threaten. The hazard remains physical. The reception becomes cultural. The microphone changed hands again, and with it the perceived meaning of the evidence.

This is why polarisation matters strategically. It distorts risk perception not merely by making some citizens believe the wrong thing, but by changing how institutions themselves behave. Staff become more cautious in what they publish. Ministers learn to reward certainty that flatters their coalition and to punish candour that complicates it. Agencies begin to anticipate backlash and trim their own language before anyone else does it for them. A planning memo becomes a political weapon. A hazard map becomes a loyalty test. The concept to name here is identity capture. It refers to the process by which a technical issue is drawn into broader contests of belonging and power until the institutional conditions for neutral assessment begin to erode.

The strategic costs are concrete. Delayed action is one. When every risk communication is read first as partisan positioning, agencies and elected officials lose time arguing over legitimacy before they can argue over substance. Staff intimidation is another. Scientists, analysts, emergency planners, and local officials who produce unwelcome findings may face harassment, reputational attack, or career penalties, formal or informal. Policy whiplash is a third. One administration builds a framework, another dismantles it to mark ideological distance, a third restores part of it under new branding, and the operating system beneath the public continues to weaken. These are not cultural side effects. They are governance failures with direct operational consequences.

Delayed action is perhaps the easiest to see and therefore easiest to underestimate. A flood defence is postponed because the underlying risk assessment has become politically toxic. A managed retreat discussion is deferred because leaders fear it will be framed as surrender. A wildfire prevention measure stalls because acknowledging the hazard now signals membership in the wrong camp. A climate-health advisory is watered down to avoid media attack. None of this requires open denial of the underlying risk. Delay often works through quieter mechanisms: language softened, report release postponed, consultation narrowed, or spending reframed until the policy has lost its force. The system continues to operate, but the pace of risk response is now set partly by identity conflict rather than by hazard logic.

Staff intimidation is more corrosive than public debate usually recognises. Institutions depend on people being willing to write plainly, model carefully, report unwelcome trends, and defend uncomfortable conclusions. When those acts begin to attract threats, organised pressure, or political retaliation, the damage is not limited to the individual case. Colleagues observe what happened and learn. Future drafts become more cautious. Internal dissent becomes quieter. Senior officials choose safer wording. Recruitment becomes harder in roles already demanding and underpaid. Analytical integrity is not lost only when data are falsified. It is also lost when the internal cost of telling the truth rises high enough that candour becomes professionally irrational.

The United States offers a clear example of this general dynamic, not because it is unique, but because its institutional scale makes the pattern easy to see. Climate assessments, disaster preparedness language, land management, and environmental regulation have at times become loaded symbols in wider partisan struggle. The result has not only been disagreement over policy ends. It has included pressure on the framing of evidence itself. Personnel change, terminology shifts, delayed publications, and abrupt reversals of federal emphasis all illustrate how technical institutions can be drawn into a conflict over identity in which evidence is judged less by method than by who seems to benefit from it politically. The relevance here is not to litigate every episode. It is to show how quickly a technical domain can become a partisan marker with operational consequences.

A second example can be seen in Australia, where bushfire risk, emissions debate, land management, and climate communication have repeatedly been pulled into identity-laden political contests. The point is not that disagreement in a democracy is pathological. Disagreement is normal. The point is that once the issue becomes a badge of belonging, the institutions charged with hazard assessment face a harder task. Their message competes not only with uncertainty or limited attention but with the emotional utility of simplification. A technical explanation that contains caveat, complexity, or time lag is then disadvantaged compared with a blunt narrative that flatters an existing political self-understanding. Governance becomes noisier precisely where it needs procedural steadiness.

This is why institutions need protection for analytical integrity that does not itself become fuel for culture war. The temptation, when expertise is politicised, is to answer heat with more heat: to defend science by turning every technical dispute into a moral referendum on enlightenment versus ignorance. That move can be emotionally satisfying. It is often strategically clumsy. It further fuses the evidence to identity and invites citizens already suspicious of elite institutions to hear technical advice as a declaration of social contempt. The better path is harder. It requires agencies to defend method, transparency, role clarity, and revision discipline without converting those defences into partisan self-performance. Analytical integrity is strongest when it is procedural, boring, and stubborn.

Procedure matters here because it shifts the ground of legitimacy. An institution can say, in effect, that its claim deserves attention not because its staff are morally superior or politically neutral in some mystical sense, but because the evidence was gathered in a specified way, reviewed through a specified process, revised when new information emerged, and published within a defined institutional remit. That does not make the agency immune to attack. It gives the public a basis for trust other than charisma or ideological affinity. It also helps staff. Where method and role are clear, employees can point outward to a structure larger than themselves. Where they are not, every analyst becomes individually exposed to political projection.

Still, procedure is not enough on its own. Institutions also need role discipline. A scientific office should not try to win a culture war it cannot control. A political executive should not force technical agencies to carry rhetorical burdens that belong to elected leadership. Emergency managers should not be left to defend national climate policy every time they issue an operational warning. The more these roles blur, the more each office inherits conflicts it cannot resolve. Polarisation thrives in blurred space. Clear roles do not abolish it, but they make it easier to keep the essential functions intact: measure, warn, update, decide, explain.

A fair limitation must be acknowledged. Polarisation is not always produced by cynical manipulation from above, and public scepticism toward expertise is not always irrational. Institutions do fail. Experts do sometimes overstate, dismiss local knowledge, or communicate in ways that sound contemptuous, vague, or politically naive. A government that invokes expertise selectively, hiding behind it when convenient and ignoring it when not, teaches the public to treat expert claims as another factional tool. The answer, therefore, cannot be a simple demand for deference. It must include institutional self-discipline: consistent use of evidence, visible correction of error, and willingness to explain uncertainty without making expertise sound like an unquestionable priesthood.

This limitation strengthens the main argument because it clarifies what protection of analytical integrity should mean. It should not mean insulation from all criticism. It should mean insulation from improper interference with method, publication, staffing, and internal professional judgement. Democracies are entitled to contest policy choices. They are less safe when political actors learn they can selectively intimidate the people who produce the information on which serious policy choice depends. A minister may reject a recommendation. That is politics. It becomes a different matter when the minister punishes the production or publication of the evidence itself because the evidence disrupts the preferred story. At that point the state is not merely debating policy. It is degrading one of the mechanisms by which policy remains tethered to reality.

Policy whiplash deserves special attention because it is where polarisation becomes a direct administrative cost. A system built around long horizons, infrastructure exposure, or risk communication cannot function well if each electoral change rewrites the basic narrative of danger and preparedness.

Staff lose continuity. Local governments hesitate to align their own plans with national guidance that may soon be repudiated. Private actors postpone investment. Publics grow cynical because every new framework appears temporary and performative. The hazard, meanwhile, does not pause for ideological turnover. What looks like democratic alternation from the outside can feel like institutional vertigo from within. Strategic trust is hard to build in a building whose signs are replaced every few years.

This is one reason governance mechanics matter more than partisan detail here. The question is not which party or tendency is right in any given national argument. The question is what institutional arrangements make it harder for identity conflict to corrupt the public handling of risk. Independent publication schedules help. Protected technical roles help. Clear separation between evidence assessment and political choice helps. Local consultation that reveals how risks are lived, rather than only how they are abstractly modelled, helps too because it broadens the social base of legitimacy. When technical institutions are seen to listen downward as well as report upward, their claims are less easily caricatured as elite imposition.

The role of local knowledge is therefore not ornamental. It can provide a counterweight to identity capture because it grounds risk communication in recognisable conditions: the road that floods, the clinic that loses power, the field that no longer yields, the settlement that needs another evacuation. None of this replaces expert analysis. It helps anchor it socially. A population asked to accept a risk assessment is more likely to treat it seriously if the assessment sounds connected to the world people know rather than only to remote models or ideological camps. Institutions protecting analytical integrity should not retreat into technocratic hauteur. They should widen the evidentiary conversation without surrendering method.

A useful counter-example lies in agencies or jurisdictions that have managed to keep at least some climate-related functions comparatively insulated from extreme identity conflict by anchoring communication in practical service continuity rather than abstract grandstanding. Where warning systems, infrastructure assessments, local planning, and emergency management remain tied to concrete outcomes - keeping roads open, reducing fire exposure, protecting water systems, improving evacuation timing - public debate may still be heated, but the institutions retain more room to work. The lesson is not that politics disappears when communication becomes practical. It is that practical framing can sometimes prevent expertise from being swallowed whole by symbolic contest.

The microphone changed hands in Brazil in 2019, in Australia during fire debates, in the United States across changing administrations, and in many quieter disputes where a hazard map or warning bulletin suddenly became a proxy war over belonging. That is why the real question is not whether climate issues will continue to polarise; many will. The harder question is whether institutions can build enough procedural strength, role clarity, and social grounding that when the microphone changes hands again, the evidence does not have to change its character with it just to remain audible.

Media ecosystems and crisis amplification

In August 2023, as wildfire threatened Yellowknife in Canada's Northwest Territories, official guidance had to compete not only with smoke, distance, and transport constraints, but with the speed of rumour. The place was clear. The date is easy to verify. A message about road conditions, fuel availability, shelter space, or airlift priority no longer travelled through one public channel at a measured pace. It moved through phones, neighbourhood groups, screenshots, anxious family calls, clipped interviews, partisan commentary, and fragments repeated with growing certainty as they travelled further from their source. The signal multiplied.

That multiplication is the right place to begin because modern media do not merely report crises. They shape their operational environment. A hazard becomes socially real through narratives of urgency, reassurance, blame, and explanation, and those narratives now move through systems designed to reward speed, emotion, identity, and repetition. Institutions that still treat communication as something done after operational decisions are settled are therefore governing only half the emergency. The other half is already circulating, accelerating public reaction before the formal apparatus has finished its briefing note. The signal multiplied again, and with it the range of behaviours that officials now had to manage rather than merely observe.

This is why crisis amplification should be treated as an ordinary condition of emergency management rather than as a spectacular exception. Panic, complacency, and scapegoating are not random public moods descending from nowhere. They are social products intensified by media structures that privilege dramatic certainty over qualified guidance. A false claim that a route is closed can crowd a second route within minutes. A minimising narrative that officials are exaggerating can delay evacuation until options narrow sharply. A rumour that one group is receiving preferential treatment in shelters can provoke anger against strangers who are themselves under acute stress. The concept to name here is communicative load. It matters because the emergency system must now carry not only people, fuel, and information, but also the behavioural effects of messages moving through channels it does not control.

The first danger is panic, though that word is often used too loosely. Most people do not become irrational in crises. They respond to what seems timely, plausible, and socially confirmed. Media ecosystems can sharpen that plausibility very quickly. A single image of a crowded petrol station, an out-of-context clip of an official, or a confidently worded post from someone apparently "on the ground" can trigger real behaviour before any correction arrives. Queue formation, early departures, over-purchasing, route congestion, and self-directed shelter movement often emerge not because the public is foolish, but because the information environment has shortened the interval in which patient verification feels possible. Emergency management therefore has to anticipate behavioural cascades generated by communication, not merely by the hazard itself.

The second danger is complacency, which is often less dramatic and therefore more politically convenient. Crisis communication is usually discussed in relation to fear, but media ecosystems are just as capable of dissolving urgency as of inflating it. A selective narrative that past warnings proved excessive, that officials are speaking from political motive, or that "real locals" are unconcerned can drain compliance even when formal forecasts and operational indicators justify early action. This is especially dangerous in evolving emergencies where the cost of leaving too early feels immediate and visible, while the cost of leaving too late remains hypothetical until it is catastrophic. Complacency thrives when scepticism is socially rewarded and when institutions have not built enough prior trust to make qualified warning sound like competence rather than melodrama.

Scapegoating is the third danger, and often the most corrosive over time. When stress rises, media ecosystems are well suited to supplying culprits faster than institutions can supply explanation. Evacuees are blamed for crowding shelters, newcomers for straining services, utilities for every hardship, local authorities for national failures, national authorities for local bottlenecks, minorities or outsiders for visible disorder, and experts for uncertainty itself. Some criticism is warranted in any serious crisis. Scapegoating is different. It simplifies causation by attaching structural strain to a person or group convenient to resent. Once that narrative settles, it becomes harder for institutions to communicate trade-offs honestly because any admission of limit is immediately recoded as proof of betrayal by the chosen villain.

This is why misinformation should be treated not as an embarrassing by-product of crisis but as a predictable operational condition. A state that regards false or misleading information as an occasional nuisance to be corrected when time permits will keep arriving too late. The better view is plainer: misinformation is now part of the hazard field. It shapes movement, demand, compliance, trust, and hostility. It does so whether or not it originates from organised disinformation, partisan media, careless commentary, frightened relatives, or ordinary misunderstanding. The source matters for long-term accountability. Operationally, the immediate question is how the information will alter behaviour before the next official update appears.

A useful example comes from evacuation contexts, where rumours about route closure or shelter conditions can directly undermine compliance. Imagine an evacuation order issued with one primary route recommended and two secondary options held in reserve. Soon a message spreads through local groups claiming the primary route is already blocked and that officials are hiding the fact to avoid embarrassment. Some families leave earlier than advised, others switch to less suitable roads, and some delay departure because they now doubt the official picture altogether. Traffic patterns change. Fuel demand shifts. Police and transport managers must now respond to a behavioural disturbance created by communication rather than by flame, wind, or flood alone. The hazard has not merely advanced physically. It has been socially amplified.

A named case illustrating the broader principle is Hurricane Katrina in 2005. Much of the media coverage revealed genuine institutional failure and should not be dismissed. Yet the event also showed how quickly fragmented reporting, rumour, partial images, and repeated frames of disorder could shape public perception and political response beyond the immediate physical situation. The point is not that media created the crisis. They influenced how it was interpreted, escalated, and remembered. In later emergencies, authorities across many jurisdictions learned - sometimes imperfectly - that communication cannot wait until facts are complete because perception is already organising action while fact-finding continues. That lesson remains only partially absorbed.

A second example can be drawn from wildfire seasons in North America and Australia, where social media rumours about road status, looting, shelter conditions, air quality, and official motive have repeatedly complicated operations. Again, the aim is not to exaggerate isolated falsehoods into omnipotent forces. Most crisis outcomes are still driven primarily by material conditions and institutional competence. The analytical point is narrower and more important: false or distorted information reliably exploits the same pressure points that emergencies already strain - fuel supply, route choice, trust in shelter systems, perceptions of fairness, and willingness to comply with inconvenient guidance. Media ecosystems amplify whatever those systems are least prepared to stabilise.

This is why communication must be incorporated into emergency design at the same level as logistics, transport, and medical coordination. A resilient system does not only ask where buses, generators, or shelter beds are. It asks where the public will look for the next trusted update, who can speak for route conditions, how local leaders and journalists will be briefed, what false claims are most likely to emerge, which communities receive information late or through distorted intermediaries, and how often official messages must appear in order to remain behaviourally relevant. Communication is not the glossy narration of operational competence. It is one of the means by which competence is exercised.

The role of journalists deserves careful treatment here. Media ecosystems are not reducible to falsehood. Professional journalism often performs essential public functions during crisis: verifying claims, exposing institutional weakness, amplifying urgent local realities, and compelling candour from officials who might otherwise drift into vague reassurance. The problem arises when the wider media environment collapses distinctions between verified reporting, selective framing, emotional testimony, political theatre, and algorithmic amplification. In such conditions, the most operationally useful information can be drowned out by what is most engaging, identity-confirming, or anger-producing. Institutions need serious journalism. They cannot assume that serious journalism alone will define the information environment in which they must act.

Still, caution is necessary. It would be a mistake to imagine that better official messaging can neutralise all amplification effects. Some rumours will spread no matter what. Some publics will remain suspicious for reasons rooted in prior institutional failure, not present communication alone. Some media actors will deliberately intensify controversy because controversy is their business model or political strategy. The claim here is not that communication mastery can abolish the problem. It is that institutions worsen the problem when they speak too slowly, too defensively, too generally, or without enough role clarity to establish which channel should govern immediate behaviour. A weak official signal invites stronger unofficial ones.

This is where credibility and repetition matter more than single-message brilliance. During crisis, people often decide which information to trust not by comparing every claim analytically, but by asking which source has previously proved timely, specific, and honest under pressure. That is why boring consistency beats rhetorical flourish. The official channel that updates route conditions at expected intervals, admits uncertainty, corrects itself plainly, and distinguishes operational instruction from broader commentary can gradually become behaviourally dominant even in a noisy environment. The channel that appears only to deny rumours or defend reputation usually will not. A system that treats misinformation as predictable prepares its own rhythm in advance.

Participatory trust also matters here, even before the next section addresses it directly. Communities are more likely to ignore destructive rumours when they already know how official systems work, who local intermediaries are, what constraints shelters face, how route updates are issued, and why guidance may change. That familiarity cannot be improvised in the hour of danger. It must be built beforehand through ordinary encounters, public explanation, and planning processes that allow people to see their own realities reflected in official design. The signal multiplies less destructively when the public has some prior reason to recognise its legitimate source.

The deeper strategic point is that media ecosystems redistribute time. They compress the interval between perception and action while lengthening the institutional consequences of communication failure. A rumour can alter behaviour immediately. A correction may arrive later, after the route is crowded, the fuel is gone, the shelter is distrusted, or the scapegoat has already been chosen. Emergency managers who understand this do not ask only whether a message is accurate. They ask whether it is early enough, specific enough, repeated enough, and anchored enough in existing trust to compete in the real communications environment rather than the one imagined by official procedure.

The signal multiplied in Yellowknife, as it does in nearly every modern emergency. That is no longer a side story. It is part of the emergency itself. The harder question is not whether misinformation will appear; it will. The harder question is whether institutions can build communication systems strong enough that when the next false route closure, shelter rumour, or scapegoating narrative begins to circulate, the public still knows which signal is meant to govern behaviour - and hears it before the amplification becomes one more force the emergency has turned against them.

Participatory planning and legitimacy

In November 2021, after intense rainfall and flooding disrupted transport corridors and communities in British Columbia, the public saw roads closed, shelves thinned, and official coordination tested under pressure. The place was western Canada. The date is easy to verify. Yet beneath the visible emergency sat a quieter question that every serious planning system eventually faces: who had been asked, beforehand, how the region actually worked when roads failed, supply routes narrowed, and formal assumptions met lived constraint? The map is not enough.

That matters because risk planning is still too often imagined as a technical exercise completed by experts and then delivered to the public as an instruction set. Experts are indispensable. Models, engineering judgments, hazard maps, and institutional planning all matter. But plans built at a distance from the people expected to live under them routinely miss critical facts: which road is the true local bottleneck, which clinic cannot function on paper staffing assumptions, which households cannot evacuate without transport support, which shelters are socially unusable for some residents, which warnings arrive too late to matter because people need more time than officials assume. The map is not enough again, because the real test of planning is not whether it looks coherent in a briefing deck. It is whether it holds when ordinary lives press back against it.

This is the central argument. Legitimacy is produced, not assumed. Public institutions do not possess durable compliance simply because they hold legal authority or technical expertise. They build it through repeated encounters in which people can see that plans reflect actual constraints, trade-offs, and values rather than administrative convenience alone. The concept to name here is participatory legitimacy. It refers not to endless consultation as a moral ritual, but to the structured inclusion of affected knowledge and affected priorities in the making of risk decisions. The implication is exacting: a plan that is technically sound but socially tone-deaf may still fail strategically because people will not comply, or will comply too late, when disruption comes.

This does not mean participation is a substitute for decision. Governments still have to choose, prioritise, and sometimes impose burdens people do not welcome. The value of participatory planning lies elsewhere. It improves plans by revealing what formal systems routinely overlook. A flood map may show inundation zones accurately and still miss the neighbourhood routes families actually use to leave. A shelter plan may allocate beds efficiently and still ignore medication storage, disability access, gendered safety concerns, or the reality that people will not abandon animals, relatives, or work equipment as easily as the plan imagines. An emergency communications strategy may identify official channels and still fail to notice that a given community receives information through local radio, school staff, elders, faith leaders, or neighbourhood groups long before it trusts a distant ministry. Participation, then, is not sentiment. It is operational correction.

A named example from New Zealand helps clarify the point. In Christchurch after the earthquakes that began in 2010 and intensified in 2011, public trust in planning and recovery did not depend only on engineering competence or central authority. It also depended on whether communities believed that rebuilding, zoning, relocation, and infrastructure decisions reflected the realities of place rather than only top-down administrative logic. The details of that recovery were contested, as large recoveries always are. That is precisely why the example matters. It shows that legitimacy is not a soft extra added after the technical plan. It is part of whether the technical plan can be executed without accumulating resentment, resistance, and later non-compliance.

This is especially important under climate-linked risk because many of the relevant decisions are distributive and morally loaded. Whose area gets protected first? Which road is maintained to a higher standard? Which settlement receives early warning resources? Which neighbourhood is marked for retreat, redesign, or more intrusive regulation? These are not questions that data alone can settle, even when data are strong. They require judgments about fairness, identity, memory, economic life, and acceptable loss. Institutions that treat such judgments as though they were merely technical often find that the public does not reject planning because it is irrational, but because it can recognise when value choices have been hidden inside technical language. Participation does not abolish conflict over such choices. It makes the conflict more governable because people can at least see what is being weighed and why.

This is where consultation improves planning most clearly. It can reveal local bottlenecks, overlooked vulnerabilities, informal care networks, cultural obligations, housing realities, transport limits, and distrust patterns that central systems either do not see or do not know how to rank. A municipality may discover that a designated evacuation route is unusable for a large share of older residents. A regional authority may learn that a proposed shelter is technically adequate and socially impossible because of past experiences, location, or community conflict. A utility may find that its restoration priorities underestimate the importance of a local service hub. A housing authority may learn that relocation plans assume mobility, documentation, or savings people do not possess. These are not small details. They are the difference between a compliant public and a bewildered one.

A second example lies in flood and coastal adaptation planning, where participatory processes have often exposed the limits of purely model-driven approaches. Communities asked to accept stronger building restrictions, altered insurance expectations, or eventual retreat from vulnerable land rarely respond well to maps and timelines alone. They ask, reasonably, who will pay, who moves first, what happens to equity, whether tenants are being protected as well as owners, and how the place they know is being valued beyond its risk category. Where these questions are ignored, planning may proceed formally while legitimacy drains away. Where they are engaged seriously, the plan may still be difficult, but it is more likely to survive the transition from document to practice. The map is not enough because the map does not carry memory or obligation on its own.

Still, participation is not free. It consumes time, staff attention, political patience, and administrative capacity that many institutions do not possess in abundance. Meetings must be organised, materials translated, feedback processed, trade-offs explained, and expectations managed. Front-line officials who are already stretched by service delivery may now be asked to design processes of listening as well. Consultation can also be captured by the loudest voices, overrepresent the already organised, or harden opposition if authorities invite participation without making clear what is genuinely open to change. This is a real limitation, and it must be faced honestly. The defence of participatory planning is not a defence of endless process. It is a defence of the right amount of structured involvement to make plans more realistic and more politically durable than closed design usually permits.

That limitation sharpens the core point. Because administrative time is scarce, institutions need to know what kind of participation actually builds capacity. Not every forum does. The most useful mechanisms tend to be those that illuminate constraints and values relevant to operational success. Which populations would struggle to comply with this instruction and why? Which services appear substitutable on paper but are not substitutable in practice? Which trade-offs will be seen as unfair unless explained early? Which informal networks help the state and which will route around it if trust is weak? Participation is productive when it answers such questions. It becomes hollow when it asks communities to perform consent without influence or to provide information the system has no intention of using.

This is why legitimacy should be treated as a strategic asset rather than a moral atmosphere. It affects compliance during disruption in ways that are visible and measurable. People evacuate sooner when they believe the route and shelter system have been designed with some knowledge of their actual circumstances. Residents tolerate temporary hardship more readily when they recognise their own concerns in the logic of the plan. Communities share information more openly when previous consultations produced changes they could see. Local leaders amplify official warnings when they trust that the official system listens back. Legitimacy, here, is not abstract approval. It is the practical willingness to cooperate with inconvenient decisions under stress.

The reverse is equally true. A legitimacy-poor system may still issue lawful orders, publish technically elegant plans, and stage impressive briefings. Yet during disruption it discovers that households delay leaving, routes clog unpredictably, shelters go underused, local actors produce rival guidance, and rumours travel faster than official messages. What appears from above as irrational behaviour is often the accumulated consequence of plans made without sufficient regard for how people actually live. The system then pays twice: once for the weak plan, and again for the compliance problem the weak plan helped create. Participation, in that sense, is not merely inclusion. It is a way of reducing the later costs of governability failure.

A fair counter-example should be acknowledged. Some highly technical systems do perform well with limited public participation because the operational variables are narrower, the infrastructure more centralised, and the choices less ethically distributed. Not every decision requires broad consultation. Not every community process improves the result. In acute crisis, authorities also need enough decisional clarity to act without waiting for consensus that will never come. The argument here is not that every plan should emerge from participatory abundance. It is that where compliance, equity, local knowledge, and differentiated burden are central to success, legitimacy must be built beforehand or improvised later at much higher cost.

This matters for state capacity because participatory planning alters the relationship between formal authority and lived complexity. A capable state is not only one that can command. It is one that can discover enough about society in advance that command does not have to do all the work once conditions worsen. Participation broadens institutional sight. It reveals dependency chains, emotional fault lines, cultural obligations, informal infrastructures, and hidden burdens that hazard models alone will not disclose. That makes the state less blind and, often, less brittle. Capacity is therefore not diminished by all listening. In the right form, it is enlarged by it.

The politics of expertise enters here too, but differently from the previous section. Expertise is not undermined when it is asked to meet local knowledge. It is tested, corrected, and sometimes strengthened. The danger lies not in engagement but in the false binary between expert judgement and public experience. Plans fail when authorities assume only one of these counts. A flood engineer may know the water. Residents may know the route people actually take when the road closes. A health planner may understand throughput. Community workers may understand why some households will avoid a shelter altogether. Strategic planning demands both kinds of knowledge, ranked and combined rather than romantically equated.

The map is not enough, but neither is the meeting room by itself. That is the discipline participatory planning requires. It must feed into budgets, route design, warning systems, shelter standards, local staffing, and public explanation. It must also end somewhere, with decisions that remain difficult and sometimes unpopular. The real test is what follows. Did consultation alter the plan? Did the plan then perform better when disrupted life pressed against it? Did communities recognise themselves in the logic of the response even when they did not get everything they wanted? Those are the measures that matter.

The harder question, then, is not whether public participation is normatively attractive. It is whether institutions that claim to plan for severe risk can still afford to treat legitimacy as though it appears automatically at the moment of command, rather than as something produced patiently beforehand - one route, one shelter, one trade-off, one difficult conversation at a time.

Chapter 10

Adaptation Governance and Coordination Failures

What fails when everyone is responsible and no one can move first? That question reaches the centre of adaptation governance faster than a catalogue of agencies ever could. Climate risk is often described as though the main challenge were scientific uncertainty or political denial. Those matter. Yet many adaptation failures occur after the risk is already recognised and before any serious dispute over its existence needs to be resolved. A road washes out repeatedly, a drainage system proves undersized, a vulnerable neighbourhood remains in place, a monitoring system produces warning after warning, and still the governing machine stalls. The file changes hands.

That hand-off is the quiet drama of modern administration. One office owns the asset, another maintains it, a third sets the standards, a fourth holds the data, a fifth controls the money, and a sixth is blamed when disruption becomes visible to the public. Each can plausibly say the issue lies partly elsewhere. None is entirely wrong. The result is not necessarily negligence in the crude sense. It is a design problem disguised as an implementation problem. Institutions built to divide labour cleanly can struggle badly when the risk sits in the seams between them. The file changes hands again, and the cost of that movement is measured in delay, duplication, and the deadening ease of blame.

This is why coordination deserves to be treated not as a managerial virtue word, but as a strategic variable. Adaptation rarely fails because a single office refuses to act in isolation. It fails because action depends on several organisations sharing mandates, standards, information, timelines, and authority in combinations they were not built to sustain under pressure. An infrastructure ministry may recognise a vulnerability while local government bears the maintenance burden. A municipal planner may want revision while national codes still embed older assumptions. A monitoring unit may detect worsening conditions while legal rights over the data, or uncertainty about who can authorise a response, slows movement until the signal becomes another archived warning. The public sees drift. Inside the system, the drift is often distributed and therefore hard to arrest.

The central tension, then, is between specialisation and governability. Specialisation is necessary. No serious state can function without divided responsibility, technical standards, data rules, procurement controls, and layered accountability. But the same structures that produce order in calm periods can produce paralysis under climate stress if they make cross-boundary action too costly, too ambiguous, or too politically dangerous. Adaptation is especially vulnerable because it often lacks the drama that forces emergency hierarchy into place. The issue may be urgent in strategic terms and still not urgent enough in political terms to override routine fragmentation. What looks like institutional maturity can become institutional drag.

That drag matters because adaptation changes the value of time. A slow revision to a building standard, a delayed code update, an unresolved dispute over maintenance responsibility, or a classification rule that limits data sharing can all appear tolerable for a while. Under changing climatic conditions, however, these quiet delays accumulate exposure. The road built to one assumption remains in service under another. The procurement template still buys for yesterday's conditions. The standard remains lawful after it has ceased to be prudent. Governance is not only what governments decide. It is the rate at which they can revise the rules embedded in ordinary decisions before crisis turns those rules into liabilities.

The stakes therefore extend beyond administrative neatness. They concern the credibility of the state itself. When adaptation stalls in the gaps between offices, people do not experience a subtle coordination problem. They experience repeated flood damage, contradictory instructions, delayed maintenance, warnings without action, and the sense that public authority exists mainly as an argument over ownership of the problem. Confidence erodes in small increments. Local officials become more cautious, not less, because they fear owning consequences without owning tools. National ministries become more performative because visible concern is easier to produce than structural change. Over time, the state can become highly articulate about adaptation and surprisingly weak at carrying it through.

The path ahead follows that weakness where it actually forms. It begins in the gaps between mandates, where overlapping jurisdictions and unclear responsibility create the delays and duplications that later masquerade as fate. It then settles on standards, codes, and procurement rules, those slow-moving instruments through which old assumptions about climate stability keep governing long after they should have been revised. From there the focus shifts to data governance and decision rights, where preparedness is often blocked not by ignorance but by uncertainty over who may act on what is already known, before arriving at accountability across multiple levels of government, where honest reporting and prudent risk-taking can be punished by the very systems that claim to demand them.

What binds these themes is a simple but difficult proposition: adaptation is not mainly a question of inventing more policy language. It is a question of rearranging authority, incentives, and revision pathways so that institutions can act across boundaries without waiting for a perfect alignment that never quite arrives. This is why the chapter stays close to governance mechanics. The real obstacles are often mundane and therefore politically underestimated - maintenance lines, code committees, procurement templates, data classifications, reporting chains, municipal liabilities, national ownership without local support. Such details can sound bloodless until one notices that they decide whether a known risk becomes a managed transition or another case study in preventable exposure.

The file remains the image worth keeping. A warning enters the system. It is read, forwarded, reviewed, recategorised, and held. One office has concern without funds. Another has funds without mandate. A third has authority without local knowledge. A fourth has the data and cannot share it cleanly. A fifth knows the weakness and has learned from experience not to move too early lest the blame later settle there. That is how adaptation fails without anyone needing to deny the hazard. The harder question, then, is not whether governments can name climate risk. It is whether they can build institutions in which the file stops moving long enough for someone with the right authority, information, and protection to act before the next warning becomes only one more document in the chain.

Mandates, silos, and coordination costs

In August 2005, Hurricane Katrina struck the Gulf Coast of the United States and exposed failures that quickly became legible not only as failures of infrastructure or leadership, but as failures of coordination across public systems that each possessed part of the problem and none of the whole. The place was New Orleans and far beyond it. The date is easy to verify. Responsibility did not vanish. It multiplied. Federal, state, local, military, emergency, transport, and welfare institutions all had formal roles, and still the gaps between them proved decisive. The file moved.

That movement is the right place to begin because adaptation often fails not where no institution exists, but where too many exist in arrangements that reward caution, partial ownership, and procedural self-protection. A flood defence may sit under one budget line, land-use control under another, maintenance under a third, emergency access under a fourth, and political blame under all of them at once. Each office can describe its remit with perfect sincerity. Each can also treat the decisive action as slightly beyond its authority, slightly underfunded, or slightly dependent on another actor moving first. The file moved again, and the risk deepened while the organisational chart remained formally intact.

This is why coordination failure should be understood as a design problem rather than as a moral disappointment. Institutions are built to divide labour. That is not a flaw. Specialisation allows expertise, routine, and legal clarity. Yet climate adaptation places unusual pressure on systems built around sectoral boundaries because the relevant risks are cross-cutting, slow-moving, and often difficult to fit into any one office's core incentives. The concept to name here is fragmented responsibility. It matters because when responsibility is fragmented in the wrong way, delays, duplication, and blame shifting do not arise accidentally. They are the predictable outputs of the structure itself. The implication is severe: many adaptation failures are not failures to coordinate in the colloquial sense. They are failures produced by institutions whose rules make coordination costly, risky, and politically unrewarding.

One can see this most clearly where jurisdictions overlap. A national government may own a strategic road, rail line, or coastal asset while local authorities manage surrounding drainage, emergency access, land-use pressures, and day-to-day service disruption when the asset falters. The national level sees scale, network importance, and capital value. The local level sees potholes, blocked culverts, school bus disruption, ambulance delays, and residents demanding answers. If the maintenance burden or adaptation need grows, each side has strong reasons to say the other should move first. The centre may argue that local upkeep and permitting have been poor. The municipality may reply that it cannot redesign around an asset it does not own and cannot afford to defend. The result is not simply disagreement. It is a system that converts known exposure into repeated negotiation. The file moved, and with it the cost of inaction.

A useful example lies in transport and flood management systems where nationally important infrastructure depends on local drainage or access conditions. A rail corridor or trunk road may be funded and owned at national level, but repeated service failures may begin with culverts, adjacent land management, feeder roads, or drainage maintenance that sit with local or regional bodies. The public experiences one failing system. The state experiences several partially aligned ones. This matters because adaptation needs often surface first through disruption at the local edge: road closures, flooded underpasses, inaccessible stations, schools cut off, repair crews delayed. Local officials know the pain points. National authorities often control the capital decisions. When the connection between those two realities is weak, climate risk is translated into institutional delay rather than corrective action.

A second example can be found in coastal protection and port access, where ownership, environmental permitting, municipal land use, and emergency planning frequently sit in different hands. A port authority may seek adaptation works to preserve access. Environmental agencies may govern the permitting pathway. Municipalities may face the consequences of altered drainage, traffic, or shoreline effects. National ministries may control strategic funding. None of these actors is irrational to defend its own mandate. The difficulty is that the system does not naturally reward one of them for bearing the coordination cost on behalf of the whole. Meetings multiply. Assessments are commissioned. Partial works begin. Meanwhile, the hazard continues to test the seam between institutions harder than any one institution tests it internally.

This is where duplication often enters. When coordination is weak, agencies frequently respond not by doing nothing, but by creating parallel efforts. Multiple risk maps are produced because datasets, procurement rules, or institutional prestige differ. Separate consultation processes are run with the same communities because each office has its own statutory obligations. Emergency planning, maintenance planning, and capital planning all assess the same vulnerable asset through different templates and incompatible timelines. The public sees repetition and often assumes laziness. The deeper problem is misalignment. Duplication becomes a survival strategy inside fragmented systems because each office trusts its own process more than the possibility that another office's work will actually carry operational weight when decisions are made.

Blame shifting follows naturally from this structure and should not be mistaken for mere bad character. If responsibility is diffuse and political punishment is sharp, institutions learn to protect themselves. A local authority may report vulnerability carefully but avoid direct ownership of a recommendation it cannot fund. A ministry may delay explicit commitment until regional partners have formally agreed to implementation steps. A utility or infrastructure operator may produce technically sound warnings while framing action as contingent on regulatory decisions that move more slowly. In such settings, adaptation stalls not because no one can see the problem, but because moving first increases exposure to blame if the cross-agency system still fails. Risk aversion migrates from the hazard field into the administrative field.

This is why the reader should diagnose coordination costs concretely rather than rhetorically. Coordination is not free goodwill floating above institutions. It has price, time, and political risk attached to it. It consumes staff hours, legal review, meeting cycles, procurement work, and reputational exposure. Someone must absorb that cost. In badly designed systems, no one is clearly authorised or rewarded to do so. The sharper diagnostic questions are therefore practical. Who has formal authority to convene the relevant actors? Who can compel or fund joint action? Who carries the downside if an early warning is escalated and later judged excessive? Who benefits if the issue remains framed as someone else's remit? Which office bears the pain of public failure most directly, and does it possess the tools to reduce that failure? Such questions reveal coordination failure as an institutional pattern, not a vague lament about bureaucratic culture.

Still, caution is necessary. Not all overlap is dysfunctional. Some redundancy in mandates can be protective. It may create backup channels, multiple warning routes, or checks against reckless action by one office. Likewise, specialised silos are not always signs of failure. Technical depth requires boundaries. The argument here is not that adaptation governance should be flattened into one commanding centre. Many systems would become worse, not better, if every function were centralised. The stronger claim is narrower. Coordination becomes pathological when no level of government or agency can act proportionately on cross-cutting risk without either trespassing on another's remit or accepting political and fiscal exposure far beyond its own formal powers.

A fair counter-example lies in systems where clear lead-agency arrangements, standing coordination bodies, or shared funding mechanisms reduce these costs enough to make cross-boundary action more routine. Some flood management regimes, civil protection systems, and transport authorities perform better precisely because they recognise that ownership, maintenance, warning, and community consequence do not sit neatly in one office and have built institutional machinery to bridge the gap. That machinery does not eliminate conflict. It lowers the transaction cost of acting before failure becomes public scandal. This matters because it shows fragmentation is not fate. It is governed through design choices, some of which are better than others.

The human consequence of poor design should not be lost behind the administrative language. When mandates are unclear and silos strong, residents are the ones who live inside the delay. They drive the repeatedly flooded road. They lose school days, clinic access, work time, and confidence in warning systems. They watch the same infrastructure fail under a chorus of competing explanations from offices that all sound partly right and collectively absent. Trust erodes through repetition. The adaptation state then begins to appear as a relay of disclaimers rather than a structure capable of learning. That is strategically dangerous because once communities cease to believe that reporting a problem will trigger coherent action, the flow of honest local warning can narrow too.

This is one reason municipal experience matters so much in diagnosing coordination failure. Local officials often see the cross-sector reality first because they deal with residents, service interruptions, emergency access, and the reputational consequences of visible dysfunction. Yet they frequently possess the weakest formal tools. National ministries may have the money and ownership rights; regional bodies may have regulatory powers; specialist agencies may hold the technical standards; private operators may control key assets. The municipality is left to improvise at the point where all these systems meet daily life. Adaptation governance then becomes a test of whether the state can hear the level at which consequences are experienced, rather than only the level at which authority is most concentrated.

The politics of "joined-up government" often obscures more than it clarifies here. It sounds admirable and usually is. But it risks reducing a structural problem to a personality one, as though better relationships or more energetic leadership would solve the matter by themselves. Relationships help. Leadership matters. Neither substitutes for institutional design. If budgets remain separate, standards contradictory, incentives misaligned, and legal duties unclear, then exhortations to coordinate will merely produce more heroic improvisation by overburdened staff. The strategic value of the diagnosis offered here is that it redirects attention from motivational rhetoric to the mechanics of authority, funding, and revision.

This is why adaptation governance should be judged not only by what risks are named, but by what happens once those risks cross an organisational boundary. Does the system slow? Does it duplicate? Does it wait for formal consensus? Does the warning move upward while the tools remain downward? Does ownership sit at one level and maintenance pain at another? Does early reporting make institutions safer or more exposed? Answers to these questions tell us more about preparedness than any general adaptation strategy can. They reveal whether the system is built to absorb complexity or only to narrate it after the fact.

The file moved after Katrina, and after countless less famous disruptions where no single office quite owned the whole problem and every office owned enough of it to delay the rest. That is the unresolved friction at the heart of adaptation governance. If the next vulnerable asset is owned nationally, maintained locally, regulated regionally, monitored by a specialist body, and politically felt in one neighbourhood first, who exactly has both the authority and the incentive to stop the file moving long enough to act before the hazard becomes another lesson in how well-designed silos can still fail together?

Standards, codes, and the pace of revision

In June 2021, when the condominium tower at Surfside in Florida collapsed, public attention fixed on the horror of sudden structural failure and the grief that followed. The place was Surfside. The date is easy to verify. The causes and responsibilities in that case were specific, and they should not be casually folded into a climate argument they cannot bear on their own. Yet the event remains instructive for a different reason. It reminded governments, regulators, and the public that the built environment does not only fail through dramatic design error or exceptional negligence. It can also fail through slower institutional habits: postponed revision, accumulated assumptions, and standards that continue to govern conditions they no longer describe well. The rulebook stays open.

That image of the rulebook matters because standards are among the quietest instruments of power in modern states. Building codes, design standards, procurement templates, maintenance schedules, and approval pathways rarely excite public imagination. They lack the drama of emergency response and the symbolic force of grand policy announcements. Yet they shape what gets built, what gets bought, what risks are treated as normal, and what forms of prudence remain officially legible. When standards assume a world of relatively stable rainfall, temperature range, flood recurrence, ground reliability, or coastal behaviour, those assumptions do not remain trapped in technical documents. They travel into schools, depots, bridges, housing blocks, drainage systems, and evacuation routes. The rulebook stays open again, and what looks like ordinary compliance may already be a form of strategic drift.

This is why governance often changes through standards more decisively than through speeches. A minister can announce adaptation priorities. A national strategy can name resilience as a central aim. A city can publish a hazard assessment that is thoughtful, sober, and widely praised. None of this matters much if procurement rules still reward the cheapest near-term bid under outdated design assumptions, if building codes still encode yesterday's climate, or if infrastructure guidance still treats exceptional weather as statistically external rather than operationally central. The concept to name here is embedded governance. Standards govern because they decide what counts as ordinary design, acceptable cost, defensible delay, and prudent maintenance without requiring each choice to be fought at full political volume.

The difficulty, of course, is that standards move slowly for reasons that are not foolish. They are meant to be stable. Industry needs predictability. Public authorities need rules that can be applied consistently. Designers, builders, insurers, and local governments all need some confidence that the requirements they are working to will not be revised impulsively under every gust of public anxiety. That stability is a public good. It becomes a public hazard when it hardens into administrative inertia. Climate adaptation brings precisely this problem into view. The issue is not that standards exist. The issue is that standards often revise more slowly than the conditions they are supposed to organise, and much more slowly than the consequences of being wrong can accumulate.

One can see this most clearly in the difference between event recognition and rule revision. A flood occurs. Reports note that drainage capacity, siting assumptions, or elevation requirements may be out of date. Engineers, planners, insurers, and local officials all know something significant has been learned. Yet turning that knowledge into a revised code or design standard is another matter entirely. Committees must meet. Evidence must be reviewed. Cost implications must be assessed. Industry actors must be consulted. Liability concerns arise. Existing projects already in design or procurement complicate the picture. Local authorities ask whether the new standard will be funded. National regulators ask whether the evidence is broad enough for general revision or still too site-specific. Each step has logic. Together they can produce a tempo mismatch in which repeated warnings are absorbed socially faster than they are revised institutionally.

A named example from Australia helps clarify the point without forcing a crude technical comparison. After successive bushfire disasters, debates over planning controls, building standards, and risk classifications have repeatedly shown how difficult it is to translate visible hazard change into revised rules that are both politically acceptable and operationally enforceable. The relevant lesson is not that one specific code revision solved the matter or that every fire-prone region requires the same answer. It is that standards are never merely technical artefacts. They are settlements among evidence, cost, legal exposure, industry practice, and public tolerance for change. When the hazard environment shifts, the standard becomes a site of political negotiation even if the language around it remains technical. Governance moves through the committee room, the specification sheet, and the procurement clause far more often than through grand doctrine.

Procurement rules are especially revealing because they show how adaptation can be undermined by administrative definitions of value. A system that buys on lowest upfront cost, narrow specification compliance, or standardised delivery time may sound prudent and professional. Under climate stress, it can reward exactly the wrong kind of competence: designs that meet the current minimum, materials selected for present affordability rather than future exposure, contracts too rigid to accommodate revised local conditions, and maintenance arrangements calibrated to a world the asset may not inhabit for long. Procurement is often imagined as neutral implementation after policy. In reality it is where priorities become economically binding. If risk revision does not enter procurement language, adaptation remains aspirational while the state continues to purchase yesterday's assumptions in orderly, auditable form.

This is why revision pathways matter as much as the content of any specific revision. A standard may be outdated, but if the institution lacks a credible route for updating it, the knowledge of obsolescence produces frustration rather than change. Who can trigger review? What evidence threshold is needed? Which offices must sign off? Are revisions periodic, event-driven, or politically discretionary? Can local conditions justify deviation, or does the standard function as a rigid national baseline? Does funding move when compliance costs rise, or are local authorities simply handed a stricter code and left to absorb the backlash? These questions determine whether standards are living instruments or slow traps. A government may know perfectly well that its code base reflects older climate assumptions and still remain practically unable to revise without reopening half the political economy of construction and infrastructure finance.

Institutional resistance follows naturally, and it should not be caricatured as stupidity or greed, though those exist in some cases. Much resistance is structural. Engineers may fear that broad revision on uncertain projections will weaken confidence in the whole standard-setting system. Municipalities may resist tougher requirements if national funding does not follow. Industry may argue, sometimes sincerely, that rapid revision imposes costs that smaller firms cannot absorb. Politicians may worry that tighter codes will make housing less affordable or infrastructure more visibly expensive within the next budget cycle. Auditors may ask why standards are changing when older designs remain formally lawful. Each of these objections can contain a real concern. The difficulty is that together they tend to reward delay while presenting delay as prudence.

This is where a disciplined reader should resist two opposite simplifications. One is the technocratic fantasy that better evidence alone will force better standards. It rarely does. Standards are revised through institutions, and institutions rank evidence alongside cost, timing, legitimacy, and enforceability. The other simplification is fatalism: the idea that because standards are political, revision will always be too slow to matter. That is also false. Some systems do revise. Some sectors do learn. Some post-disaster inquiries, insurance pressures, judicial decisions, or repeated maintenance failures do eventually push assumptions out of the codebook and into a different official normal. The strategic question is not whether revision is possible. It is which institutions make it less punishing to move early rather than after another visible failure.

A second example can be found in floodplain development and drainage design, where formal standards often outlast the conditions under which land was first approved for housing or critical service provision. Again, the point is not to adjudicate one jurisdiction or one technical threshold. It is to show how rules become political infrastructure. Once homes, schools, roads, and utilities have been built to one standard, revising that standard means more than changing a number in a table. It raises questions about stranded assets, retrospective fairness, insurance, local tax bases, and the moral status of people already living under the old rule. This is why standards are so powerful and so hard to move. They do not only govern the future. They retroactively cast judgment on the past.

Still, caution is necessary. Standards that revise too quickly or too opaquely can also damage trust. If requirements appear to change unpredictably, public and private actors may begin to see the whole system as unstable, politicised, or impossible to plan around. This matters. A standard that no one believes will hold for long ceases to coordinate effectively. The answer, therefore, cannot be permanent revisionism. It has to be disciplined revision: clear triggers, transparent reasoning, role clarity, and pathways that explain not only what has changed but why the earlier assumption no longer deserves automatic rule status. Adaptation governance fails when standards are frozen. It can also fail when revision looks arbitrary or punitive.

That balance makes standards a particularly revealing test of institutional seriousness. A state that understands adaptation only as emergency response will underinvest in these slow levers because they do not satisfy the appetite for visible action. A state that understands adaptation as governance will eventually realise that codes, procurement frameworks, and design standards are where climatic assumptions become ordinary practice. They decide whether the next school, road, warehouse, seawall, or housing scheme quietly reproduces old exposure or begins to alter it. They make the future legible in procurement language long before disaster forces it into headlines.

The politics of compliance deepens this point. Citizens and local authorities rarely interact with "adaptation governance" as an abstract field. They encounter it in permit conditions, inspection requirements, procurement restrictions, and maintenance obligations. If those rules feel detached from lived conditions, underfunded, or imposed without a credible account of why the revision matters, compliance becomes brittle. If they are explained clearly, phased with some institutional honesty, and connected to real support or real consequence, the same rules can alter behaviour quite deeply without ever becoming ideological theatre. That is one reason standards matter so much. They govern quietly when they work. They become politically explosive only when the mismatch between rule and reality has been allowed to grow too obvious to ignore.

A fair counter-example should be acknowledged. Some infrastructure and building systems are already conservative enough, or locally adapted enough, that existing standards continue to perform reasonably well despite changing conditions. Not every code base is fatally outdated. Not every revision demand is wise. Some calls for reform are opportunistic, poorly evidenced, or insensitive to cost burdens that would fall hardest on those already stretched. A serious analysis has to admit this or risk turning "revision" into a reflex rather than a judgement. Yet even this limitation returns us to the main point. What matters is not revision for its own sake. It is whether institutions can recognise when the assumptions hidden in their standards have become too stale to keep governing responsibly.

The rulebook stays open, and that is the unresolved friction this section leaves in view. Every state says, in one form or another, that it wants to adapt before disaster teaches the lesson twice. The harder question is whether it can revise the quiet instruments through which it already governs - the code clause, the procurement template, the design assumption, the maintenance interval - before those instruments continue to purchase and legalise exposure under the respectable name of compliance.

Data governance and decision rights

In September 2018, after Hurricane Florence struck the Carolinas, public agencies, utilities, insurers, transport operators, and local authorities all held pieces of the same emergency picture. The place was the south-eastern United States. The date is easy to verify. Flood extent, road closures, hospital status, shelter occupancy, power disruption, and wastewater risk did not exist as one clean dashboard shared by all who needed to act. The problem was not simply ignorance. Much was known. The harder question was who could see which part, who could use it, and who had the authority to change operations before the next consequence arrived. The dashboard was lit.

That image matters because adaptation and preparedness are often discussed as though the central challenge were information scarcity. Sometimes it is. More often the deeper difficulty lies elsewhere. Data exist, but in separate systems, under different legal rules, inside incompatible software, behind contractual restrictions, or within institutions unsure whether acting on the data would exceed their mandate. A flood model may exist without a planning trigger. A utility may hold detailed infrastructure exposure data without a clear pathway to share it. A local authority may know which residents need transport support during evacuation, while privacy rules, procurement arrangements, or poor record integration make that knowledge hard to turn into timely coordination. The dashboard was lit again, and still the governing system hesitated.

This is why data governance should be treated as a matter of authority rather than only of knowledge. The key question is not merely "what is known?" but "who can decide?" That shift changes the entire analytic frame. A monitoring system that identifies risk has limited value if the office receiving the signal cannot authorise inspection, issue warning, alter operations, spend contingency funds, or compel another institution to respond. The concept to name here is decisional legibility. It refers to the clarity with which data are linked to roles, rights, and duties inside the administrative system. The implication is severe: a state can become saturated with information and still remain strategically blind if no one is clearly empowered, obliged, and protected to act on what the information shows.

Ownership is the first obstacle. Public institutions frequently speak as though data were a shared national resource once collected. In practice data are often attached to organisational territory. An agency collects because collection justifies budget and supports its own remit. A private operator treats data as proprietary because it was expensive to gather or commercially sensitive to disclose. A local authority guards records because they contain personal information gathered under limited purpose. A ministry classifies infrastructure data because release might expose vulnerability or legal risk. None of these decisions is automatically irrational. Together they can produce a system in which preparedness fails not because data are absent, but because they are enclosed by different logics of control.

Classification intensifies this problem. Some information is restricted for obvious reasons: national security, personal privacy, commercial sensitivity, legal confidentiality, or the prevention of malicious misuse. These concerns are real and cannot be brushed aside with technological optimism. Yet classification habits can also outlast their original rationale or expand beyond what prudence requires. A map of vulnerable infrastructure, a register of at-risk buildings, a list of medically dependent residents, or a maintenance backlog assessment may be held so tightly that the people responsible for practical preparedness can access only fragments. The result is a peculiar administrative pathology. Institutions know enough to worry, not enough to coordinate, and too little about who is allowed to bridge the gap. Secrecy becomes a form of diffusion.

Privacy concerns create a different but equally serious tension. They are often treated in public argument as either sacred barriers or bureaucratic excuses. Neither treatment is adequate. Privacy protections exist because states and corporations can do real harm with badly governed personal data. But the mere existence of privacy law does not decide how preparedness should work. Hard cases arise where emergency planning depends on some knowledge of specific need: residents requiring electricity for medical equipment, households needing assisted evacuation, care facilities with fragile staffing, or neighbourhoods where language support affects whether warnings will be understood. The issue is rarely whether such data exist. It is whether institutions have built lawful, trusted, and operationally usable pathways for acting on them before crisis compresses the timeline. Privacy without preparedness can become brittle. Preparedness without privacy becomes predatory. Governance has to hold both.

A named example that illustrates the stakes is California's wildfire and utility shut-off environment, where utilities, emergency managers, public health bodies, and local governments have faced difficult questions about vulnerable customer data, outage risk, and protective action. The point is not to claim one perfect model or to reduce every difficulty to data-sharing alone. It is to show that the practical problem is not simply forecasting fire or grid stress. It is deciding who can use information about medically vulnerable populations, route constraints, or infrastructure exposure in time to coordinate warning, transport, sheltering, and backup power support. When decision rights are unclear, the most ethically sensitive data are often the hardest to translate into action just when they matter most.

A second example can be seen in flood risk and insurance systems, where property-level exposure data, municipal planning records, insurer assessments, and infrastructure condition reports often sit in different institutional compartments. A town may know which drainage assets fail repeatedly. Insurers may know which properties present recurrent claims patterns. National infrastructure agencies may know the vulnerability of adjacent transport routes or utility networks. Citizens experience one flood system. The state and market often manage several datasets with no shared operational grammar. The challenge is not merely to merge them technically. It is to decide who may combine them, who is accountable if they do not, and who carries legal or political risk if the combined picture implies costly intervention.

This is where proprietary systems become strategically significant. Digital tools promise integration, prediction, and real-time insight. They also create dependency. A public agency may procure software it does not control fully, operate on vendor terms it cannot easily alter, or store critical risk data in formats that are hard to transfer or share. A private infrastructure operator may hold key operational data inside a platform unavailable to municipal partners. A hospital, utility, and local authority may each use systems that cannot communicate cleanly without new contracts, new permissions, or workarounds too slow for emergency conditions. The public rhetoric says the system is data-driven. The operative reality is sometimes worse than paper: the data exist, but the system's legal and technical architecture prevents them from travelling where decisions must be made.

This is why turning monitoring into action depends less on dashboards than on governance pathways. A river gauge, permafrost sensor, wildfire model, heat alert system, or outage forecast only becomes strategically useful when someone knows in advance what level of signal changes which operational choice. Does the local authority gain discretion to close a road? Does the utility have to notify healthcare providers? Does the ministry release contingency funding? Does the public warning threshold change? Does the port alter traffic, or the school district alter transport, or the care system activate outreach? Monitoring without decision rights creates a dangerous illusion of readiness. It gives the appearance of sight while leaving the limbs unsure whether they are authorised to move.

The reader should therefore ask a different set of questions when assessing preparedness claims. Not only whether the state has invested in data collection, modelling, or sensor networks, but whether the route from signal to action is clear. Who receives the alert? Who interprets it? Who can contest it? Who can override routine procedure? Who bears liability if action is taken early and judged unnecessary, or taken late and judged negligent? These questions expose a fact that data rhetoric often conceals: action is a legal and political event, not a purely informational one. Where institutions fear being punished more for acting on imperfect information than for waiting until failure is undeniable, the system will rationally drift towards delay.

Still, caution is necessary. Not every barrier to data sharing is foolish, and not every call for integrated information systems is wise. Large integrated datasets can enable surveillance, mission creep, discriminatory targeting, or false confidence in imperfect models. Proprietary systems sometimes exist because public procurement has failed to build viable alternatives. Classification can protect genuinely sensitive infrastructure and personal data from harmful exposure. A serious argument must admit these realities. The point is not to worship integration. It is to insist that where preparedness depends on coordinated action, the rules governing access, use, and authority should be designed consciously rather than inherited accidentally from fragmented bureaucratic history.

A fair counter-example lies in systems where limited but well-governed data sharing works better than maximal integration. Some emergency management arrangements succeed because they define narrow, lawful pathways: specific officials can access particular vulnerability data under particular triggers; utilities and health services share defined information during outage risk; local and national bodies rely on common geospatial references without pooling every underlying dataset. These models matter because they show that the solution is not always "more data" or "one platform". Often the better answer is role-specific access combined with clear responsibility. The governance question is not how to eliminate boundaries. It is how to make boundaries permeable enough for action without making them so weak that trust and legality collapse.

Trust is central here because people and institutions share sensitive information only when they believe it will be used competently and proportionately. A community may resist vulnerability mapping if it fears stigmatisation, insurance harm, or later exclusion from services. A private operator may resist disclosure if it believes the information will be mishandled or weaponised politically. Local officials may avoid recording certain risks explicitly if previous candour brought blame without support. Trust, then, is not an ethical accessory to data governance. It is one of the preconditions for obtaining and using the data that preparedness depends upon. A system that treats trust as secondary may end up with formally impressive data architectures and impoverished practical sight.

This connects directly to accountability. If responsibility for acting on data is unclear, then after disruption each institution can say that the warning belonged elsewhere in the chain. The modeller produced the projection. The ministry funded the platform. The local authority lacked authority to act. The operator was awaiting confirmation. The regulator did not receive the relevant feed in time. The mayor was never briefed. This is not merely blame shifting after the fact. It is often the predictable result of systems built without explicit decision rights attached to specific categories of information. The question "who can decide?" is therefore also the question "who can be held to account without destroying the willingness to report and act early next time?"

The strategic point is plain enough. A state that cannot convert data into authorised action is not merely inefficient. It is vulnerable in a distinct way. It may know more than its predecessors and still fail more expensively because its knowledge raises expectations without increasing operational freedom. The public sees maps, sensors, alerts, and platforms and assumes preparedness has advanced. When disruption comes, it discovers that the warning was real but trapped: in the wrong system, under the wrong classification, behind the wrong contract, inside the wrong jurisdiction, or attached to no one's clear duty to move first. The dashboard was lit, and still the system waited.

That is the unresolved friction this section leaves in view. The next time an agency says it is closely monitoring a severe risk, the sharper question is not whether the monitoring is technically impressive. It is whether the signal has a lawful path to the people who need it, whether they can use it, and whether they are empowered to act before the evidence is neat enough to satisfy every institutional fear. If not, the state has not solved the preparedness problem. It has only measured it more clearly.

Accountability in multi-level systems

In August 2005, when Hurricane Katrina struck the Gulf Coast of the United States, one of the most enduring public questions was brutally simple: who was in charge? The place was New Orleans and far beyond it. The date is easy to verify. Yet the practical answer was never simple. Federal agencies, state authorities, municipal officials, emergency managers, police, engineers, transport departments, and welfare systems all possessed part of the picture and part of the mandate. Responsibility did not disappear. It dispersed. The finger points somewhere.

That image is useful because accountability in multi-level systems is often discussed as though its main difficulty were moral weakness: too little courage, too little honesty, too much evasion. Those things exist. They are not the whole problem. Modern states govern through layered authority. National governments fund and regulate. Regional bodies coordinate and supervise. Municipal systems deliver, improvise, and absorb first contact with failure. During climate-linked stress, this layering can be a strength. It spreads burden, allows local knowledge to matter, and prevents every problem from waiting on a distant capital. It can also make responsibility harder to read, especially when the same infrastructure, service, or population is governed through different mandates at different levels. The finger points somewhere again, and often finds several plausible targets at once.

This is why accountability must be examined as a design question rather than a post-mortem ritual. A state can possess many accountable offices and still fail to produce accountability for the thing that matters most: whether someone with the right authority, information, and protection acted early enough to reduce known risk. The concept to name here is dispersed accountability. It refers to a condition in which formal responsibility exists at several levels, but the practical ability to assign credit, blame, or corrective duty becomes muddy because action depended on interlocking decisions no single office fully controlled. The implication is severe: without careful design, multi-level governance can punish visible failure after the fact while failing to reward the slower, quieter work of preparedness that might have reduced failure beforehand.

This matters because after-action blame is not a neutral cleansing mechanism. It teaches institutions what kinds of behaviour are safest. If officials learn that visible crisis performance is what will be judged, then they will optimise for crisis performance. If they learn that raising awkward risks early, asking for resources before disaster, or documenting institutional fragility will expose them to criticism without protection, many will become more cautious in exactly the wrong way. They will report less bluntly, escalate later, hedge more, and avoid stepping beyond narrow remit even when they can see the larger danger. A blame system designed to enforce responsibility can therefore degrade responsibility by teaching actors that honesty and initiative are personally expensive.

One can see this especially clearly in municipal government. Municipalities often sit closest to exposure. They maintain local roads, drainage, emergency shelters, housing enforcement, schools, public health links, and the ordinary interfaces through which disruption becomes visible to citizens. Yet they frequently possess the weakest fiscal margins and the least political cover. A municipality may know that a culvert system is undersized, that a school doubles as an unsuitable shelter, that backup power is inadequate, or that evacuation support for care-dependent residents is thinner than official plans imply. If earlier candour produced rebuke, if requests upward were slow or humiliating, or if local leaders were later blamed for conditions they had repeatedly flagged, the incentive to speak plainly next time narrows. The finger points downwards very easily in such systems because the local level is where the public pain becomes legible.

A named example from the COVID-19 period is instructive here, even though the hazard was epidemiological rather than climatic. In many countries and federated systems, national, regional, and municipal actors were judged harshly after waves of hospital strain, procurement failure, school disruption, and uneven local enforcement. Some criticism was deserved. Yet the episode also revealed how readily multi-level systems can punish actors for decisions made within constraints they did not control. Local officials were blamed for scarce protective equipment they did not procure, hospitals for staffing shortages shaped by national systems, and regional authorities for communication failures partly created by central political messaging. The lesson for adaptation governance is not that blame should disappear. It is that multi-level accountability must be able to distinguish failure of judgment from failure of delegated capacity.

A second example lies in wildfire and flood management, where municipalities may be held publicly responsible for visible damage even when land-use rules, insurance regimes, infrastructure ownership, and emergency funding are shared with regional and national authorities. A town may be criticised for permitting housing in hazardous land while operating inside broader legal and fiscal structures that pushed affordable development into exactly those areas. A regional authority may be condemned for a delayed evacuation order when transport, warning thresholds, and shelter capacity depend on national systems. Such cases do not eliminate local responsibility. They complicate it. Serious accountability requires tracing who had power to decide, who had duty to warn, who had money to act, and who had the authority to revise the relevant rule before the event, not merely who stood closest to the microphone after the damage.

This is why accountability that rewards learning and preparedness matters strategically. A system oriented only around crisis performance teaches the wrong lesson. It asks whether officials appeared decisive, visible, and compassionate after the event, rather than whether they maintained reporting channels, revised plans, secured contingency contracts, protected technical staffing, or escalated known vulnerabilities early enough to matter. These quieter acts are less photogenic and often less politically rewarded. They are nonetheless among the strongest indicators of state capacity. A capable accountability regime therefore does not stop at asking who failed under pressure. It asks who prepared honestly before pressure, who reported risk clearly, who updated standards, who corrected weak assumptions, and whether the institutions that did those things were strengthened or sidelined by the experience.

This is difficult because preparedness often requires responsible risk taking by public officials. An emergency manager who orders evacuation early may later be accused of alarmism if the worst outcome does not materialise. A municipal engineer who insists on expensive preventive works may be criticised for overspending if several quiet years follow. A regional administrator who raises concern about weak care continuity or housing risk may be labelled pessimistic, obstructive, or politically inconvenient. If the accountability culture punishes such anticipatory action more sharply than it punishes delayed recognition, rational officials will drift towards passivity dressed up as procedural caution. The finger points somewhere in advance too, and people inside the system notice exactly where.

Still, caution is necessary. The argument for less punitive and more learning-oriented accountability cannot become an excuse for vagueness, self-protection, or bureaucratic immunity. Public institutions hold real power. They make consequential choices. Citizens are entitled to know who failed and why. Some officials do hide behind complexity that they themselves helped perpetuate. Some organisations exploit ambiguity to avoid scrutiny. A serious account must admit this. The point is not to replace accountability with therapeutic understanding. It is to design accountability that distinguishes negligent concealment from candid reporting, reckless improvisation from prudent early action, and genuine incapacity from mere blame avoidance. Those distinctions are hard. They are also essential.

A fair counter-example lies in systems where inquiry, audit, and review have managed - at least in some cases - to improve preparedness rather than merely assign guilt. Aviation safety offers the clearest model, though the domains differ. When near misses, procedural failures, and incident reports are treated as inputs to system correction rather than only as opportunities for individual condemnation, institutions learn faster. Disaster governance can never mirror aviation exactly; hazards are broader, politics is rougher, and accountability is more public. Yet the contrast reveals something important. Where honest reporting is protected and linked to corrective authority, institutions become better at surfacing weakness before it produces spectacle. Where every report is read mainly as material for later punishment, silence becomes rational.

The multi-level character of adaptation makes this even harder because different tiers of government are judged by different clocks and audiences. National leaders are often rewarded for broad commitments, fiscal announcements, and symbolic command. Regional authorities are judged on coordination and visible operational steadiness. Municipal actors are judged on whether streets flood, schools close, shelters work, and local services hold. These are not the same forms of accountability, and they do not naturally reinforce one another. National actors may prefer vague strategic language. Municipal actors may need blunt operational candour. Regional actors may be trapped between them. When failure occurs, each level can plausibly point to the other's missing piece. The result is not the absence of accountability but its fragmentation across scales that do not add up cleanly.

This fragmentation is one reason the reader should ask sharper questions after any major disruption. Not only who appeared on television, who apologised, or whose jurisdiction formally included the damaged area. More importantly: who had the authority to revise the standard or plan beforehand? Who received the early warnings? Who had discretion to spend? Who documented the vulnerability? Who ignored, delayed, or diluted the signal? Who would have been punished for acting early and being wrong? Who was punished for speaking clearly? Such questions move us beyond theatre and closer to design. They reveal whether the system is built to produce candour and action, or only to narrate failure with increasing sophistication after the fact.

This prepares the ground for later international and Arctic governance because the same logic scales upward. As authority stretches across jurisdictions, reporting cultures, funding systems, and political constituencies, the temptation to assign blame to the nearest visible actor grows stronger even while actual causation and corrective power become more dispersed. If states cannot manage accountability well within their own layered systems, they will fare even worse where multiple sovereigns, indigenous authorities, private operators, and international bodies share pieces of the same risk field. The local inquiry already contains the seed of the larger diplomatic problem.

The finger points somewhere after every disaster. That is inevitable. The deeper question is whether accountability systems can be built so that, before the next disruption, people are rewarded for reporting plainly, revising early, and preparing in ways the public may barely notice until something goes wrong elsewhere. If not, then multi-level governance will keep producing a familiar pattern: eloquent after-action blame, cautious before-action silence, and a state that becomes very good at locating responsibility only once the worst proof has already arrived.

Chapter 11

Arctic Governance and Strategic Competition

What counts as presence in a place where distance is punishing, weather can turn procedure into improvisation, and every promise of authority is tested by whether the lights stay on? That question reaches deeper into Arctic politics than the more theatrical language of flags, patrols, and maps usually allows. Sovereignty in the Arctic is never only a matter of legal claim or military symbolism. It is judged in the lived grammar of service: whether aircraft arrive, whether communications hold, whether infrastructure is repaired, whether regulations are enforced consistently, whether search and rescue is real enough to trust, and whether residents experience the state as a durable organiser of life rather than an intermittent visitor. The signal must hold.

That is why the Arctic refuses the old comfort of separating hard power from ordinary governance. A repeated service outage in a remote settlement may look small from a distant capital and politically large to the people living through it. A failed resupply, a damaged runway, a delayed medevac, or a regulatory lapse in extraction oversight can register simultaneously as logistical weakness, legitimacy loss, and strategic message. External actors read such moments too. They do not merely see a local administrative difficulty. They see evidence about state capacity in a region where credibility depends on routine competence more than on declaratory ambition. The signal must hold again, because in such settings reliability is itself a form of geopolitical speech.

This is the chapter's governing tension. Arctic governance is asked to sustain continuity under conditions that reward fragmentation. Strategic competition sharpens the pressure, but much of the real stress appears in quieter forms: dispersed populations, thin service networks, expensive maintenance, fragile logistics, strained relations between central governments and local communities, and institutional arrangements that can be disrupted more easily than they can be rebuilt. The Arctic therefore should not be imagined as a blank stage onto which rivalry simply arrives from outside. It is already a demanding administrative environment. Competition matters because it alters the terms under which ordinary governance is judged, funded, and politicised.

That judgment matters for more than symbolism. A state that cannot keep infrastructure functional, coordinate safety services, or regulate development credibly will struggle to persuade either residents or outsiders that its wider Arctic posture deserves confidence. This is not because legitimacy and strategy are separate goods awkwardly forced together. It is because they are entangled. Service provision builds belonging. Belonging strengthens the credibility of regulation and long-term planning. Credible regulation affects how firms, communities, and external observers assess risk. Risk perception, in turn, influences investment, compliance, and the willingness to cooperate across borders or institutions. Weak governance is therefore never just weak governance. In the Arctic it becomes a signal about what kind of order is really being sustained at the edge.

The stakes rise further when development enters the picture. Resource extraction and transport are often narrated as opportunities, sometimes as necessities, occasionally as existential threats. None of those frames is sufficient on its own. The harder question is how risk is distributed when environmental uncertainty meets industrial ambition. Who carries the cost if infrastructure degrades faster than expected, if spill response proves slower than promised, if transport corridors become less reliable, or if local communities absorb harms that central treasuries and private firms can price more comfortably than residents can live with? Development is therefore never simply economic. It is a test of whether governance can assign exposure and responsibility in ways that remain politically durable when conditions become harsher than the prospectus implied.

This is why scientific cooperation has retained such importance even when wider relations are strained. In the Arctic, technical exchange is not an optional courtesy layered over geopolitics. It often underpins safety, monitoring, and the reduction of dangerous misperception. Shared research, joint observation, data exchange, and pragmatic contact can help maintain a common picture of conditions in places where misunderstanding carries real costs. Yet these same mechanisms remain vulnerable to politicisation, suspension, or narrowing when strategic relations worsen. The result is a peculiar fragility: some of the most valuable forms of cooperation are located precisely in the domains most easily treated as secondary when rivalry sharpens. The loss is not only diplomatic. It is operational.

The path ahead follows that entanglement where it is most revealing. It begins with sovereignty as something performed through service provision rather than only announced through claim, then turns to resource development as a stress test of governance design rather than a referendum on extraction in the abstract. From there the argument moves into scientific cooperation, where technical exchange can still steady the system even when political trust is thin, before settling on the rules and institutional continuities that make Arctic order more than a collection of episodic bargains. The movement is deliberate: from visible presence, to distributed risk, to technical trust, to the continuity of institutions that must survive periods of strain if any of the rest is to endure.

What binds these themes is continuity. Arctic governance depends unusually heavily on institutions that can persist across weather, distance, fiscal cycles, and political temperature. When continuity weakens, every other problem grows sharper. Service interruptions feel more political. Development disputes become harder to arbitrate. Scientific channels become easier to politicise. Rules and norms that once made behaviour predictable begin to look provisional. This is why institutional continuity should be treated not as bureaucratic comfort but as a strategic resource. It gives actors a reason to plan, report, comply, and coordinate on the assumption that today's arrangements will not be casually discarded tomorrow.

That does not mean continuity is always virtuous. Some institutions carry inherited inequities, rigidities, or exclusions that deserve challenge. But if they are suspended, narrowed, or repurposed under strain, the costs travel widely. People learn that what looked stable was contingent. Partners recalculate. Local trust thins. Data sharing slows. The practical burden of improvisation rises. Strategic competition then works not only through overt rivalry but through the erosion of the quiet frameworks that once kept misunderstanding, duplication, and opportunism within limits. Governance becomes louder and less steady at once.

The signal remains the image worth keeping. A distress call, a weather warning, a regulatory notice, a service interruption, a scientific update, a ministerial promise - each asks whether Arctic governance is still capable of carrying meaning across distance without distortion or collapse. The harder question, then, is not whether the Arctic will matter strategically; it already does. It is whether the states and institutions claiming to govern it can make their presence credible through enough continuity, service, and restraint that when the next test arrives, the signal that reaches residents and rivals alike is one of durable capacity rather than intermittent control.

Sovereignty, presence, and service provision

In August 2016, the cruise ship *Crystal Serenity* transited the Northwest Passage under intense public and governmental attention. The vessel's route was real, the date is easy to verify, and the place was the Canadian Arctic. Much of the commentary turned to symbolism: opening waters, strategic access, renewed attention to the North. But the more serious question lay elsewhere. If something went wrong, who would find the ship, reach the injured, manage evacuation, coordinate ports, sustain communications, and show that authority in Arctic space meant more than a line on a chart? The signal had to hold.

That question goes to the centre of Arctic sovereignty because sovereignty in such regions is not expressed chiefly through rhetoric. It is expressed through services. Search and rescue, runway maintenance, fuel delivery, policing, communications reliability, regulatory inspection, air traffic support, medical evacuation, and the simple continuity of public administration all perform the state in ways residents and outsiders can actually test. A government may assert jurisdiction in law, and law matters profoundly. Yet in the Arctic, law that cannot travel through weather, distance, and infrastructure quickly starts to look thinner than its own language suggests. The signal had to hold again, this time not in the courtroom or diplomatic note, but in the ordinary machinery of life.

This is the central argument. State presence in the Arctic is credible to the degree that it remains serviceable. The concept to name here is functional sovereignty. It refers to authority made tangible through the reliable provision of protection, regulation, infrastructure, and care across difficult space. The implication is severe: governance capacity is not merely a domestic administrative concern that sits beneath "real" strategy. It is itself a form of strategic signalling. Residents read it as evidence of whether the state values their lives on equal terms with those lived further south. External actors read it as evidence of whether the state can actually organise the territory it claims to govern. In this setting, a delayed medevac, an unreliable runway, or repeated communications failure can speak more loudly than a ministerial declaration.

Search and rescue reveals this with particular clarity because it is one of the few functions that combines humanitarian necessity, technical capacity, and geopolitical meaning in a single act. A distress call at sea or in a remote settlement does not ask the state to perform symbolic presence. It asks whether the aircraft, vessel, runway, fuel, weather intelligence, communications, and coordination system are real enough to shorten the distance between alarm and help. That is why search and rescue should not be romanticised as Arctic heroism. It is public administration under unforgiving conditions. A rescue system that works imperfectly but honestly, with clear limits and real preparation, strengthens legitimacy because it demonstrates a state willing to translate claim into effort. A rescue system that exists mainly in plans and ceremonial language weakens legitimacy far beyond the immediate incident.

Infrastructure belongs in the same frame because it decides whether service can be routine rather than improvised. Roads, ports, airstrips, storage depots, communications links, power systems, and public buildings are sometimes discussed as development assets on one page and sovereignty assets on another. In the Arctic those categories collapse into one another. A functioning runway is a health asset, a logistics asset, a political asset, and a sovereignty asset all at once. A stable communications network is not only economic infrastructure. It is part of whether government instruction reaches communities, whether emergency management is credible, whether schools and clinics can function, and whether local frustrations accumulate into a sense of administrative abandonment. States often speak of presence as though it were mainly visible. In remote areas, presence is often audible only through the continuing hum of functioning systems.

Regulation matters for the same reason. External observers often imagine Arctic governance through naval patrols, shipping routes, or resource licensing. Yet regulatory capacity is where sovereignty becomes disciplined rather than merely declaratory. A state that can inspect, enforce safety regimes, coordinate environmental response, and maintain predictable administrative standards is showing that it does not merely claim a region. It governs it. A state that licences development but cannot monitor it, promises safety but cannot inspect it, or proclaims standards that lapse at the edge of distance is sending a different signal. Regulation, then, is not the dry tail end of strategy. It is one of the ways the state demonstrates that its writ travels beyond capital-city imagination into the conditions where private actors, local communities, and foreign observers can test it.

The political impact of repeated service outages in remote areas captures this argument in miniature. A power failure or telecoms outage in a major metropolitan centre is serious; it is also more likely to be treated as a temporary interruption in a system fundamentally assumed to work. In a remote northern community, repeated outages can acquire a different meaning. They become evidence. They suggest that the state is willing to accept a lower standard of continuity for some populations than for others, that repair logistics remain fragile, that redundancy is thin, and that official language about commitment to the North may exceed operational reality. Residents do not need a theory of sovereignty to interpret that. They experience it every time a clinic struggles, a school closes, a flight is disrupted, or a household cannot rely on the basic systems that make ordinary life governable.

Canada offers a useful example because its Arctic and northern discourse has long intertwined sovereignty with presence, yet the actual burden of sustaining service across remote communities remains extremely demanding. The point is not to single out Canada as uniquely weak or uniquely strong. It is to note that the political language of northern stewardship is continually tested by infrastructure reliability, medevac reach, telecommunications, housing conditions, and the everyday endurance of public systems. When those systems falter repeatedly, the problem is not merely one of inconvenience. It cuts into the state's credibility with its own northern residents while also complicating the message sent to other Arctic actors about what Canada can organise, fund, and sustain over time.

A second example can be seen in Greenland, where questions of infrastructure, service provision, and state presence carry significance not only for residents but for wider strategic conversations involving Denmark, Greenlandic self-government, and external powers attentive to Arctic opportunity and access. Again, the point is not to collapse complex constitutional and political questions into one functional test. It is to show that remote service reliability has political meaning beyond the service itself. When transport systems are uncertain, when settlement viability depends on fragile links, or when administrative capability appears thin relative to strategic rhetoric, the discussion about sovereignty and future governance acquires a sharper edge. External actors notice where the system strains. So do the people living inside it.

This is why legitimacy for residents and credibility for external actors cannot be separated neatly. A state may try to project confidence abroad while tolerating chronic under-service at home, but in the Arctic these two audiences observe some of the same indicators. External actors study rescue reach, infrastructure condition, regulatory follow-through, and local dissatisfaction because these are practical clues about how much real capacity sits behind formal claim. Residents study the same things because they reveal whether official attention translates into durable support. The resulting linkage is strategically important. A government that cannot sustain the confidence of northern communities may still retain legal authority, but it will find strategic messaging harder to make believable. Sovereignty without trusted service grows brittle from within.

Still, caution is necessary. The argument here is not that every outage, delay, or administrative failure signals geopolitical weakness in some grand sense. Arctic conditions are hard. Weather closes routes. Infrastructure ages. Repair costs are high. No state can abolish distance, and no responsible analysis should pretend otherwise. Some service disruptions are exactly what they first appear to be: difficult but bounded operational failures in a harsh environment. The stronger claim is narrower and more useful. When disruptions are repeated, when response is systematically slower or thinner than official narratives imply, and when communities learn to expect fragility rather than continuity, the cumulative political meaning changes. Governance then begins to communicate something its own rhetoric may not intend.

That limitation matters because it prevents the argument from collapsing into melodrama. States do not prove sovereignty by achieving perfection. They prove it by showing seriousness in the face of imperfection: maintaining infrastructure with discipline, communicating honestly about limits, revising systems after failure, and treating remote residents as central political subjects rather than as scenic proof of national identity. A fair counter-example lies in systems that, though stretched, demonstrate this seriousness through sustained investment, better rescue integration, stronger communications planning, or transparent correction after breakdown. Such cases matter because they show that Arctic capacity is not an impossible ideal. It is a demanding administrative achievement.

The phrase "state presence" is often used too loosely, as though presence were just there or not there. Presence is graded. It can be ceremonial, intermittent, extractive, protective, or relational. In the Arctic, the most durable form is relational because it joins symbolic claim to repeated service in ways people can verify. A flag can be seen once. A functioning airstrip, dependable regulation, or timely rescue is encountered again and again. Over time those encounters shape the public meaning of the state more deeply than almost any speech about northern destiny can. Strategic competition sharpens this truth, but it does not create it. The state was already being measured in these terms before rivalry intensified.

This is also why governance capacity should be read as signalling even when no one intends it primarily that way. When a remote community experiences repeated power failure, delayed repairs, or unstable transport, the message sent is not controlled by the ministry responsible. The message is inferred by those affected and by those watching. When search and rescue systems function credibly under strain, the same is true. Signals are generated by performance, not by declaration alone. That is the harder political fact. States do not get to decide unilaterally what their Arctic presence means. Their systems help decide for them.

The signal had to hold for *Crystal Serenity*, and it has to hold for the far less visible routines of Arctic life: the clinic flight, the fuel shipment, the weather warning, the school reopening after outage, the inspection visit that actually occurs. The unresolved question is not whether sovereignty in the Arctic will continue to be asserted. It will. The sharper question is whether the states making those assertions can sustain enough service, enough continuity, and enough honest competence that their own residents - and those beyond them - will continue to hear those assertions as evidence of governance rather than as words travelling alone across too much distance.

Resource development and risk externalities

In May 2020, the diesel tank collapse near Norilsk in northern Siberia forced into public view a truth that Arctic politics often prefers to keep abstract. The operator was NTEK. The place was the Russian Arctic. The date is easy to verify. What failed was not only a storage system. What failed was a distribution of risk that had allowed environmental uncertainty, infrastructure fragility, and public consequence to drift apart until they met in one expensive and highly visible event. The bill travelled outward.

That outward movement is the right place to begin because Arctic resource development is rarely governed by the firm alone, the state alone, or the community alone. Extraction and transport decisions produce value in one set of books and exposure in several others. A project may be commercially rational on its own terms while depending on public rescue capacity, public infrastructure, public legitimacy, and public tolerance for uncertainty it does not fully carry. A firm can insure part of the risk, engineer against part of it, and price part of it. The rest moves: into municipal services, into regional transport systems, into state liability, into community trust, into the political meaning of repeated disruption. The bill travelled again, and with it travelled the real question of governance.

This is the central argument. Arctic "development" should be read not as a neutral label for investment activity, but as an institutional stress test. The concept matters because development in difficult environments reveals how a governing system allocates burden before crisis and after it. It shows who bears uncertainty, who defines acceptable exposure, who pays for monitoring, who pays for emergency response, who carries the social and ecological residue of failure, and who gets to treat certain forms of risk as background conditions rather than existential concerns. The implication is exacting: the strategic problem is not simply whether a project proceeds. It is how risk is distributed among firms, communities, and states, and what that distribution teaches each actor about the credibility of the larger order.

Environmental uncertainty sharpens this problem because it changes the value of confidence. Extraction, storage, shipping, and supporting infrastructure in Arctic or sub-Arctic settings have always required judgement under difficult conditions. Yet when ground reliability, coastal conditions, ice behaviour, weather patterns, and maintenance windows become less predictable, the governance burden rises. Firms may adapt through higher technical standards, revised contingency assumptions, and more expensive operating models. States may respond with stronger regulation or more public support. Communities may be asked to accept more disruption in exchange for jobs, revenues, or promises of improved infrastructure. None of these responses is inherently illegitimate. The question is whether the system makes those reallocations visible and accountable, or whether it leaves them buried beneath the language of development until failure forces a more honest ledger into view.

This is why externalities are politically decisive. They are often treated as a technical concept in economics, a way of naming costs not fully captured in price. In Arctic governance the term must be taken more seriously. An externality is not only a cost left off a balance sheet. It is a transfer of vulnerability. A new transport route may increase pressure on search and rescue systems funded publicly. A port expansion may alter local risk without giving local authorities equivalent fiscal means. A mine may rely on housing, roads, and medical support whose strain is borne by communities and municipalities. A pipeline, storage site, or shipping corridor may depend on environmental monitoring and spill response architecture maintained partly by the state. Once these transfers are recognised, development no longer looks like a discrete project. It looks like a test of whether governance can distribute gains and liabilities in a way that remains durable when conditions worsen.

The Norilsk case remains instructive because it made this transfer visible in concentrated form. Whatever the precise causal chain one assigns to that disaster, the public meaning was broader than one failed tank. It raised questions about infrastructure integrity under changing northern conditions, about the adequacy of maintenance and oversight, about who ultimately pays when private assets generate public emergency, and about what sort of state presence exists in regions where industrial activity is strategically valued but public systems remain stretched by distance and climate. That is the larger lesson. A high-profile industrial accident can expose years of quietly socialised risk. Once exposed, the issue becomes political because no actor can any longer pretend that development's hazards were neatly contained within the firm.

A second example lies in offshore and coastal hydrocarbon development across Arctic-adjacent regions, where extraction decisions are inseparable from shipping corridors, rescue obligations, environmental monitoring, and regulatory credibility. One need not adjudicate the merits of any one project to see the pattern. A firm may calculate profitability against operating cost and projected return. A coastal community may calculate the project against local employment, cultural disruption, and spill risk. A state may calculate it against revenues, sovereignty signalling, and strategic presence. These are not the same calculations, and they are not made on the same timeline. Governance becomes the mechanism that either translates among them or allows one of them to dominate while calling the result balanced. That translation is where political legitimacy is made or squandered.

This is why infrastructure deserves such close scrutiny in development debates. Arctic projects rarely arrive alone. They pull behind them roads, depots, airstrips, housing, ports, communications systems, fuel storage, inspection regimes, and emergency plans. These systems can improve local service. They can also create long-term obligations that outlast the project or exceed the original business model. If a state invests heavily to support extraction, local residents may reasonably expect that the same commitment will also secure schools, clinics, maintenance, and reliable connectivity. If the project later contracts, automation rises, or commodity conditions shift, the state may find itself defending infrastructure and expectations it can no longer justify on the same terms. Development then becomes a test not only of economic design but of whether public commitments were built around durable governance or around a temporary industrial rationale dressed up as permanent regional policy.

Still, caution is necessary. It would be a mistake to speak as though all Arctic development merely disguises exploitation or inevitable governance failure. Some projects do deliver durable public benefits, stronger infrastructure, and improved service provision. Some firms operate under serious regulatory discipline. Some states do build institutions that reduce the transfer of hidden risk. Some communities actively seek development under terms they judge acceptable or strategically useful. These realities matter. The argument here is not a moral veto. It is an analytic correction. The question is not whether development is good or bad in the abstract. It is whether the governance design makes risk legible, allocates responsibility credibly, and preserves enough institutional capacity that communities and states are not left absorbing liabilities they were never fully empowered to negotiate.

This limitation strengthens the central claim because it turns attention away from ideological camps and towards institutional mechanics. How are contingency costs assigned? Who funds emergency readiness? Who inspects and how often? Who owns transport infrastructure once built? Who carries decommissioning burdens? Which liabilities remain private, which become public by default, and which are simply left vague until the first serious failure? These are the questions that distinguish a project governed as a bounded commercial enterprise from one governed as a broad public gamble whose risks have been quietly socialised. The more uncertainty enters the environmental setting, the more consequential those mechanics become.

Trust enters here in a highly practical form. Communities asked to tolerate extraction or transport risk do not judge governance only by legal text or ministerial promise. They judge it by whether past commitments were kept, whether outages were repaired promptly, whether inspections were meaningful, whether warnings were acknowledged, whether compensation was credible, and whether public systems seemed ready for the very harms they insisted were unlikely. External actors judge similarly. Investors, insurers, neighbouring states, and regulators all watch to see whether governance systems can actually absorb the risks they licence. Development, then, becomes a signalling environment. If governance appears thin, reactive, or politically selective, the project sends one message. If it appears disciplined, transparent, and operationally serious, it sends another.

This is one reason Arctic competition cannot be separated cleanly from governance design. States often imagine that encouraging development enhances strategic presence. Sometimes it does. But presence gained through weakly governed development can prove brittle. A project that deepens local distrust, overwhelms service provision, or produces repeated emergency strain may weaken the very legitimacy it was expected to reinforce. Conversely, a system that governs development credibly may gain not only economic return but also stronger local consent, better infrastructure continuity, and greater external confidence in the state's competence. The strategic meaning of development therefore depends less on the announcement of the project than on the distribution of risk once the project is operating under real conditions.

A useful way to read this is to ask where uncertainty goes when contracts are signed and approvals are issued. Does it remain substantially with the firm, in the form of higher operating cost and enforceable liability? Does it shift to communities through housing pressure, service strain, and environmental exposure? Does it move into the state through rescue obligations, monitoring burdens, infrastructure support, and eventual cleanup or stabilisation costs? Usually it moves in all three directions. What matters is whether the movement is acknowledged and governed. Hidden transfers are politically toxic because they make future conflict feel like betrayal rather than like the open consequence of a difficult bargain.

The language of "development" can obscure all of this because it suggests motion in a positive direction without specifying who is carried and who is left holding the weight. In Arctic politics, that vagueness is especially costly. A road, port, mine, or shipping corridor can indeed be a form of development. It can also be a mechanism through which future maintenance, risk, and legitimacy burdens are loaded onto institutions already operating with little slack. The difference lies not in rhetoric but in design. Governance worthy of the name makes these burdens visible early enough that they can shape the decision, not merely the inquiry after failure.

A fair counter-example lies in cases where strong regulatory systems, public scrutiny, and negotiated local arrangements have constrained the worst forms of externalisation. These cases matter because they show that the stress test can be passed, at least provisionally. Yet even here the lesson is not comfort. It is vigilance. Passing the test once does not eliminate the need for revision as environmental conditions shift and service burdens change. Governance has to keep pace with the moving edge of uncertainty. Otherwise yesterday's well-designed arrangement becomes tomorrow's quiet transfer of risk.

The bill travelled in Norilsk, and it travels in less dramatic ways wherever extraction and transport decisions lean on public systems, local tolerance, and climatic assumptions that no longer hold as steadily as they once did. The unresolved question is not whether Arctic development should occur in some absolute sense. It is whether states, firms, and communities can build arrangements in which the gains, burdens, and duties remain aligned closely enough that when uncertainty turns expensive, development still looks like governance under pressure rather than like a test the institutions were never honestly prepared to pass.

Scientific cooperation amid rivalry

In May 2008, the five Arctic coastal states met in Ilulissat, Greenland, to state that the existing law of the sea framework remained the basis for governing the Arctic Ocean. The place was Greenland. The date is easy to verify. The meeting is often remembered as a diplomatic signal about order rather than scramble. But its deeper lesson lay in something less theatrical. Even where strategic suspicion persists, actors still need enough shared factual ground to avoid mistaking environmental change, operational hazard, or ordinary technical activity for something more dangerous than it is. The weather station still matters.

That is the right place to begin because the Arctic punishes misperception. Distances are long, infrastructure is thin, weather can narrow options quickly, and routine activity in one domain can look ambiguous from another if the common factual picture is poor. A delayed vessel, an unusual ice pattern, a damaged runway, an unstable slope, a drifting instrument array, or a shift in permafrost behaviour can all carry safety implications before they carry political ones. Scientific cooperation is often discussed as a noble ideal placed awkwardly beside harder geopolitical realities. In the Arctic it is better understood as a practical method for reducing avoidable confusion in an environment where confusion can become expensive very quickly. The weather station still matters because it is one of the places where shared reality is built.

This is the central argument. Shared research, data exchange, and joint monitoring are often most valuable precisely in technical domains that support safety rather than prestige. The concept to name here is operational cooperation. It does not require sentimental trust or broad strategic alignment. It requires enough institutional discipline to maintain common reference points where misunderstanding would otherwise raise the cost of governing, regulating, rescuing, or simply interpreting events. The implication is exacting: scientific cooperation in the Arctic is not merely about knowledge accumulation. It is part of how states reduce misperception, support safety, and preserve a minimum of practical order even when wider relations are strained.

The value of such cooperation becomes clearest when one asks what happens without it. Monitoring systems fragment. Data are delayed, narrowed, or withheld. Baselines diverge. Technical language begins to carry political suspicion. The same environmental signal may then be read differently by neighbouring states, regulatory bodies, local communities, and commercial actors. An unexpected ice condition looks like a local anomaly to one institution and a strategic pattern to another. A shared concern over coastal erosion, pollution risk, or navigation safety becomes harder to coordinate because the underlying picture is no longer jointly held. The problem is not only that less is known. It is that what is known becomes less governable because no actor can be sure which parts of the picture are mutually recognised.

This is why scientific fora matter even when relations are poor. Fora, working groups, data exchanges, joint observation programmes, and technical contacts create venues where professionals can compare methods, correct misreadings, and preserve some continuity of shared interpretation. That continuity is often more important than public rhetoric acknowledges. Safety in the Arctic depends on many small acts of common reference: agreeing what conditions have changed, what indicators matter, how monitoring is interpreted, and which technical anomalies require explanation rather than reflexive suspicion. The practical benefit lies less in idealised harmony than in reducing the chances that actors begin operating from incompatible realities.

A useful example is the long-standing role of Arctic scientific and monitoring cooperation under institutional umbrellas that have, at various times, remained more resilient than the wider political climate might suggest. The Arctic Council's working bodies, and the broader ecosystem of scientific coordination around them, have often been valuable not because they dissolved rivalry, but because they helped sustain habits of exchange on issues such as monitoring, pollution, environmental assessment, and safety-relevant knowledge. The lesson is not that such institutions are invulnerable. It is that they can provide continuity in technical domains where continuity itself has strategic value. They help preserve a shared frame even when the political room for broader agreement narrows.

A second example can be found in meteorological and sea-ice related cooperation more broadly. Weather and ice information are not diplomatic ornaments. They are operational necessities for shipping, aviation, search and rescue, community resupply, field science, and industrial safety. When states and institutions maintain compatible observation, exchange, and interpretation practices in these domains, they are doing more than supporting research. They are lowering the risk that ordinary activity becomes more dangerous through informational fragmentation. This is the quiet power of technical cooperation. It does not eliminate strategic competition. It reduces the background noise against which competition, safety, and routine governance all have to be managed.

Still, caution is necessary. Scientific cooperation is not politically innocent. It can be instrumentalised, selectively frozen, narrowed for signalling purposes, or used performatively while more substantive relations degrade. Research access can become bargaining material. Data sharing can be interrupted by sanctions, export controls, national security concerns, or domestic political pressure. Technical fora can be asked to carry diplomatic weight they were never built to sustain. A serious account must admit this. The argument is not that science floats above politics in a zone of pure reason. It is that some forms of cooperation remain worth protecting because their failure produces practical harms that extend far beyond symbolic diplomatic loss.

This vulnerability to politicisation is especially important because the very domains most useful for safety are often the easiest to undervalue until they begin to fail. Shared monitoring, standardised observation, field access, and regular professional contact can look routine and therefore expendable beside higher-profile instruments of statecraft. Yet once they are suspended, narrowed, or repurposed, the cost may appear elsewhere: weaker situational awareness, slower regulatory learning, greater suspicion about observed change, and a reduced ability to distinguish routine hazard from politically charged anomaly. In this sense, the connection to earlier discussions of monitoring and trust is direct. Institutions trust one another more realistically, not more romantically, when they can test claims against some common technical ground.

This is also why the phrase "scientific cooperation" should not be romanticised into a promise of harmony. The stronger and more defensible claim is narrower. Cooperation is often most useful where it supports practical safety and shared interpretation in conditions of persistent difference. Joint monitoring of environmental change, data exchange relevant to navigation or infrastructure reliability, and continued contact among technical experts can all help reduce the chance that rivalry becomes distorted by bad or incompatible information. These are modest goods. In the Arctic, modest goods can be strategically large because operating margins are already thin.

The reader should also resist the opposite temptation: treating all technical exchange as naive or strategically compromising. States do have legitimate concerns about dual-use knowledge, sensitive locations, critical infrastructure, and the broader political context in which cooperation occurs. Not every form of openness is wise. Not every joint project should proceed under all circumstances. The point is that blanket suspicion can degrade the very capacities states need to manage their own Arctic responsibilities safely. Where technical cooperation directly supports search and rescue, environmental understanding, monitoring continuity, or hazard interpretation, the loss of that cooperation may weaken national capability as well as regional stability.

A fair counter-example lies in situations where scientific contact has continued only in narrowed form, or where parallel national systems have partly substituted for shared platforms. Some knowledge can still be produced domestically. Some safety systems can still function under looser exchange. This matters because it prevents the argument from slipping into dependency rhetoric. States are not helpless without perfect cooperation. But the substitution is rarely clean. Parallel systems can increase cost, reduce comparability, and narrow the common factual basis on which actors understand the region. The difference between "possible" and "well-governed" is large, and it grows larger when political strain persists.

What makes scientific cooperation strategically interesting, then, is not that it resolves rivalry but that it can keep rivalry from contaminating every technical domain equally. That is a profound service in a region where technical domains matter intensely. Search and rescue, offshore safety, shipping, community supply, infrastructure planning, and environmental response all depend on some degree of stable knowledge exchange or at least compatible observational practice. If those foundations erode, states are left with more uncertainty and fewer trusted channels through which to interpret it. Strategic competition then becomes sharper not because adversaries are suddenly more aggressive, but because the practical means of ruling out error, accident, or misreading have deteriorated.

The weather station remains the right motif because it is ordinary, exposed, and indispensable. It does not make grand policy. It helps make reality mutually legible. The same is true of shared monitoring systems, technical fora, and research cooperation built around safety-relevant domains. They are not substitutes for diplomacy, nor shelters from politics. They are part of the basic architecture through which states operating in the Arctic decide whether they are inhabiting one rough common environment or several increasingly incompatible versions of it.

That is the unresolved friction this section leaves in view. If rivalry deepens and institutions narrow, which technical domains should states still fight to keep sufficiently shared that safety, monitoring, and ordinary interpretation do not collapse into mutual suspicion - and how much Arctic order can survive once the common factual ground becomes just another theatre of competition?

Rules, norms, and institutional continuity

In May 2008, at Ilulissat in Greenland, the five Arctic coastal states affirmed that the law of the sea framework remained the basis for governing the Arctic Ocean. The place is easy to verify. The date is fixed. The declaration was often read as a statement about legal order in a region attracting growing strategic attention. It was that. It was also something quieter and, in practice, more demanding: a pledge that Arctic politics would continue to rely on rules, habits of restraint, and institutional continuity rather than on improvisation under pressure. The line held.

That image matters because Arctic governance depends less on a single commanding institution than on a layered fabric of forums, norms, procedures, technical bodies, bilateral habits, and legal expectations that make behaviour more legible than the geography itself would suggest. The Arctic is vast, difficult, and expensive to govern. Distance, weather, sparse infrastructure, and uneven state capacity all increase the value of predictability. Rules tell actors what counts as ordinary conduct. Norms signal what kinds of assertion will be treated as destabilising. Institutions provide recurring places where disagreement can be channelled, narrowed, or delayed before it hardens into something more dangerous. The line held again, not because conflict vanished, but because continuity made uncertainty less explosive.

This is the central argument. Institutional continuity is not a bureaucratic luxury in the Arctic. It is a strategic resource. The concept deserves that description because continuity preserves the practical expectation that tomorrow's interaction will still occur within a recognisable framework, even if relations are strained and interests diverge. The implication is severe: when institutions are suspended, narrowed, or repurposed under geopolitical strain, the loss is not merely procedural. It alters how states, communities, firms, and technical bodies assess risk, interpret one another's moves, and prioritise scarce capacity. A meeting cancelled, a working group frozen, a norm abandoned, a reporting channel politicised - each may look modest in isolation. Collectively they can thicken ambiguity in a region where ambiguity is already costly.

This matters because Arctic governance has long relied on a peculiar balance. States compete, but they also share an interest in preventing avoidable miscalculation, maintaining safety, and preserving enough order that navigation, monitoring, extraction, environmental protection, and emergency response do not all become hostages to every diplomatic rupture. The region has never been free of power politics, nor should its institutions be romanticised as zones of pure cooperation. Their value lies elsewhere. They stabilise expectation. They make it easier to distinguish ordinary friction from exceptional provocation, routine technical contact from strategic encroachment, and difficult governance from systemic breakdown. That stabilising function becomes more visible when it begins to weaken.

The Arctic Council offers the clearest example, not because it governs everything, but because it has embodied this balance in recognisable form. For years it provided a venue where Arctic states and permanent participants could sustain practical dialogue, support working groups, and preserve a cooperative vocabulary around technical and environmental issues even when broader strategic relations were not warm. Its value was never reducible to communiques. It lay in repeated contact, procedural memory, and the maintenance of an expectation that some Arctic business would continue to be conducted in institutional time rather than only through crisis diplomacy. When such continuity narrows or stalls, the loss is felt not only in diplomatic symbolism. It is felt in the weakening of habits that once made cooperation on monitoring, safety, and shared problems more administratively normal.

A second example lies in the broader ecosystem of Arctic agreements and operational arrangements around search and rescue, scientific exchange, environmental assessment, and incident-relevant technical contact. Each arrangement may appear limited in scope. Together they help create an environment in which actors can assume that not every disagreement requires starting from zero. That is the political gift of continuity. It saves time, reduces transaction costs, and lowers the threshold at which strained relations become operationally dangerous. When institutions are repurposed for narrower signalling goals, or suspended because wider rivalry makes contact politically uncomfortable, the apparent act of firmness can produce a quieter strategic cost: the erosion of shared reference points that once helped keep competition from contaminating every domain.

This is why the distinction between suspension, narrowing, and repurposing matters. Suspension is the most visible form of rupture. Meetings stop, projects pause, contacts cool, procedural calendars are interrupted. Narrowing is subtler. Institutions remain in place but their agenda shrinks, their ambition falls, their technical work becomes more cautious, and their ability to host difficult but necessary discussion diminishes. Repurposing is subtler still. A forum survives, yet the meaning of participation changes. It becomes a vehicle for national posture, selective signalling, or reputational management rather than a stable setting for practical coordination. Each of these shifts leaves a different mark. All of them affect continuity. All of them alter how much institutional life can still cushion the region against wider geopolitical turbulence.

Still, caution is necessary. Institutional continuity is not an absolute good. Some institutions can become stale, exclusionary, or too polite to confront changing realities. Norms can preserve inequity as well as restraint. Continuity can harden into ritual if participants mistake process for problem-solving. A serious argument must admit this. The case for continuity in the Arctic is not a case for freezing every arrangement exactly as inherited. It is a case for understanding what continuity does before treating disruption as costless. When states suspend or repurpose institutions, they are not merely shedding diplomatic ornament. They are altering the operating environment in which future safety, monitoring, bargaining, and even selective cooperation will have to occur.

That is why institutional continuity should be analysed with the same discipline this book has applied to infrastructure, communication, and coordination. One should ask what function an institution performs beyond its formal mandate. Does it preserve regular contact across strained relationships? Does it anchor technical exchange needed for safety? Does it provide a procedural route for raising concern before concern becomes accusation? Does it give smaller actors or remote communities some confidence that they will still be heard within a recognisable structure rather than only after a larger power chooses to notice them? These are not sentimental questions. They are strategic diagnostics. A durable norm or forum may be worth preserving not because it solves everything, but because the vacuum left by its decay is harder to govern than the institution ever was.

The consequences of continuity loss also travel downward. Residents in Arctic regions do not necessarily follow institutional design in abstract terms. They do feel the results. When norms weaken and coordination thins, search and rescue may become harder to organise, scientific monitoring more fragmented, regulatory expectations less predictable, and development disputes more vulnerable to politicisation. Municipal and regional authorities can find themselves operating with thinner external support and less confidence that wider systems will still function as before. Firms may adapt by becoming more cautious, more extractive, or more dependent on private arrangements. Communities may infer that strategic rhetoric is outrunning practical order. Institutional continuity, then, is not only for diplomats. It conditions the reliability of the systems ordinary life increasingly depends upon.

This is where the chapter's larger concern with prioritisation under uncertainty begins to emerge. A state cannot preserve every institution, every habit of cooperation, every forum, and every technical exchange in identical form once geopolitical strain intensifies. Choices will be made. Some channels will be treated as essential, others as expendable, others as politically impossible for a time. The strategic challenge is to know which continuities matter most to retain if the wider field becomes less stable: which rules still anchor behaviour, which forums still reduce misperception, which technical ties still support safety, which norms still signal restraint in practice rather than merely in language. Prioritisation becomes serious at the moment when continuity can no longer be taken for granted.

The line held at Ilulissat because states chose, for that moment, to affirm a legal and institutional grammar rather than indulge a more chaotic one. The harder question now is not whether Arctic institutions can remain untouched by rivalry; they cannot. It is which continuities are so valuable that allowing them to fray would make the region strategically harder to read, harder to govern, and more vulnerable to the very uncertainty that all serious Arctic actors claim they wish to manage rather than multiply.

Chapter 12

Scenario Design for Feedback Risks

What kind of planning remains responsible when the future can no longer be treated as a single line extending neatly from the present? That question sits at the centre of every institution asked to govern under feedback risk. Budgets, maintenance cycles, staffing models, procurement rules, and political promises all carry quiet assumptions about tempo, sequence, and recoverability. Most of the time, those assumptions stay hidden because the world behaves well enough for them to pass as common sense. Under climate feedback risk, they begin to show. A system built for interruption may be forced to face accumulation. A plan designed for one emergency at a time may overlap. What looked prudent starts to depend on a future that no longer feels singular. The map folds.

That is why scenarios matter, and also why they are so often misunderstood. They are not crystal balls for anxious officials or literary exercises for policy retreats. They are disciplined tools for exposing the assumptions already embedded in institutions that imagine themselves as realistic. A scenario does not tell a government what will happen. It tells a government what it has quietly assumed must happen in order for its current choices to remain defensible. Once that becomes clear, the exercise stops looking speculative and starts looking administrative. The map folds again, and the stress appears not in the imagination but in the plan.

The difficulty is that institutions often want more certainty from scenarios than scenarios can honestly give. Leaders ask which future is most likely. Departments want one planning baseline. Treasuries prefer one set of numbers. Communications teams prefer one message. Yet feedback risks do not respect the bureaucratic desire for one authorised storyline. They sharpen the difference between forecasting and preparedness. Forecasts try to describe what is expected. Scenarios test what would follow if conditions moved along different pathways, at different speeds, through systems with different tolerances for strain. That distinction is not academic. It is the line between planning that rehearses trade-offs and planning that mistakes confidence for control.

This matters because the greatest institutional weakness is often not ignorance of extreme possibilities, but attachment to the most comfortable version of the future. Organisations develop preferred worlds. In those worlds, deterioration is gradual enough for budgets to adjust, cooperation remains available when needed, infrastructure degrades visibly before it fails, and warnings arrive early enough to preserve orderly choice. Such worlds may be plausible. They are also flattering. They preserve the dignity of present arrangements. Scenario work becomes valuable precisely when it interrupts that comfort without sliding into spectacle. The aim is not to terrify institutions into motion. It is to force them to look at the consequences of their own dependencies, especially where those dependencies are hidden inside schedules, staffing assumptions, mutual aid expectations, and recovery timelines that no one has seriously challenged.

The stakes are therefore practical before they are philosophical. A government that cannot distinguish scenarios from forecasts may either dismiss difficult futures as melodrama or cling to them as if they were destiny. Both errors are disabling. The first produces complacency dressed as realism. The second produces paralysis dressed as seriousness. Good scenario design avoids both traps by keeping attention on mechanisms. How would a slow erosion of reliability alter procurement, maintenance, and public trust over time? How would a rapid cascade across several systems overwhelm routines built for sequential recovery? Which decisions remain reversible, and which ones close the space for adjustment later? These are not questions about prediction. They are questions about institutional quality.

This is also why tail risks must be handled with care. Extreme outcomes exert a peculiar fascination in public life, especially when media systems reward dramatic narrative and when institutions feel accused of timidity. Yet the answer to timid planning is not theatrical planning. Tail risks deserve a place when they are anchored in plausible mechanisms rather than headlines, and when they reveal vulnerabilities in decision rights, logistics, or communication that would matter even if the full extreme never arrived. The discipline lies in refusing both denial and performance. A serious institution does not ask only whether a dramatic scenario is imaginable. It asks what chain of failure, pressure, or misjudgment would make it administratively relevant. Once that question is posed properly, the exercise becomes less cinematic and more demanding.

The chapter's movement follows that demand from method to commitment. It begins by separating scenarios from forecasts so that planning can recover its proper object: choice under uncertainty rather than certainty by other means. It then presses into the harder territory where extreme but plausible outcomes must remain in view without hijacking judgement, before turning to structured challenge as a way of testing whether plans survive contact with inconvenient realities. From there the argument arrives at the point where monitoring, authority, and action must be linked through pre-agreed pathways if uncertainty is to produce accountability rather than drift.

What unifies these concerns is the question of institutional courage. Not courage in the dramatic sense of emergency rhetoric, but in the quieter sense of allowing cherished assumptions to be tested before events test them more brutally. Red teaming, stress testing, and trigger-based planning all threaten comfortable illusions. They reveal whether mutual aid would really arrive, whether communications would really hold, whether authority is clear under load, whether thresholds that look objective in a slide deck actually commit anyone to move when they are reached. Such exercises can feel adversarial because they expose weakness before the public does. That discomfort is part of their value. It is how institutions learn whether they are planning for strain or merely planning to sound prepared.

The strategic gain is not mastery of the future. No scenario system can provide that. The gain is more modest and more important: better prioritisation under uncertainty. When organisations rehearse contrasting worlds honestly, they can begin to see which capacities matter across several futures, which commitments should be made only with exit options, which triggers need public legitimacy as well as technical precision, and which decisions should be treated as provisional rather than quietly irreversible. Scenario design becomes a bridge between imagination and governance only when it changes what institutions notice about themselves.

The map folds because the future will not lie flat for anyone. The harder question is whether institutions can use scenarios to expose their own hidden bets before those bets harden into public failure - or whether they will keep asking scenario work to deliver the one comfort it cannot honestly provide, which is the promise that uncertainty itself can be managed by turning it into a single story.

Scenarios versus forecasts

In October 1962, during the Cuban Missile Crisis, the most dangerous choices in Washington and Moscow were made under conditions of partial knowledge, compressed time, and catastrophic possibility. The date is easy to verify. The places are obvious. Yet the relevant lesson is not chiefly military. It is cognitive. Neither side possessed a stable picture of what would happen next, and any pretence of certainty would have been reckless. Decision-makers still had to choose. They could not wait for the future to become clear before acting inside it. The map had already changed.

That is why the distinction between scenarios and forecasts matters so much in public planning. A forecast tries to describe what is expected to happen. It narrows the field. It assigns weight, projects trend, and offers the comfort of a leading line. Sometimes that is exactly what institutions need. A weather forecast, a budget projection, or a demand estimate can guide routine choice well enough because the system being managed is relatively bounded. A scenario does something different. It widens the field. It does not tell an institution what future will arrive. It asks what its current assumptions would look like if several plausible futures began to press on them in different ways. The map had already changed again, and the point was no longer prediction. It was preparedness for choice.

This is the central argument. Scenarios are tools for thinking, not predictions in costume. The distinction is more than methodological tidiness. It changes how institutions treat uncertainty. When scenarios are mistaken for forecasts, two errors usually follow. One is dismissal: the scenario is rejected because it cannot prove it will happen. The other is false authority: the scenario is treated as a quasi-prediction and then blamed for alarmism or failure when events unfold differently. Both errors miss the purpose. A good scenario is not an attempt to win a wager against the future. It is a disciplined device for surfacing the hidden assumptions already embedded in plans, budgets, staffing models, and political promises.

Those hidden assumptions are often more consequential than the scenario itself. A maintenance budget may assume deterioration remains gradual enough to absorb through normal cycles. A mutual aid agreement may assume neighbouring jurisdictions will remain available when called. A supply plan may assume transport interruptions arrive one at a time. A treasury may assume contingency spending will be episodic rather than recurrent. None of these assumptions necessarily appears in the headline text of the plan. They sit inside timelines, procurement rules, reserve levels, service promises, and silent definitions of what counts as manageable disruption. Scenario work matters because it pulls those assumptions into view before events force them into public embarrassment.

This is why good scenarios improve decision quality by rehearsing trade-offs rather than by pretending to foresee outcomes. Rehearsal is the key word. Institutions rarely fail only because they lack imagination. They fail because they have not practised choosing between competing goods under pressure. A scenario that poses a "slow grind" world, in which reliability erodes gradually but repeatedly, forces different questions from a "shock cascade" world, in which several systems fail in tight sequence and recovery windows collapse. The first presses on maintenance tempo, cumulative fiscal fatigue, public patience, and the politics of incremental deterioration. The second presses on decision rights, reserve capacity, mutual aid, communication under stress, and the ability to protect essential services while several demands arrive at once. Neither world needs to be forecast as probable in order to be useful. Each can expose where the current system has bet too heavily on one tempo of change.

The phrase "slow grind" is especially valuable because it challenges one of the most common institutional reflexes: the preference for crises that declare themselves clearly. Governments are often better at mobilising for dramatic interruption than for prolonged attrition. In a slow grind planning world, roads do not necessarily fail spectacularly; they degrade more often and cost more to keep barely serviceable. Housing does not become uninhabitable in a single shock; it becomes more fragile, more expensive to insure, and less reliable for the people least able to move. Public trust is not shattered overnight; it is worn down by repetition. A scenario built around such a world helps institutions test whether their budgets, staffing, and political narratives can survive deterioration that is serious but never quite cinematic enough to command emergency powers.

The "shock cascade" planning world tests a different weakness. Here the issue is not whether institutions can endure cumulative wear, but whether they can still make decisions when assumptions about sequence collapse. An outage affects communications. Communications failure slows transport coordination. Transport disruption delays fuel delivery. Fuel delay weakens health services and shelter operations. Public messaging lags, and misinformation fills the gap. In such a world, the value of the scenario lies not in predicting a precise chain of events. It lies in asking which rights to decide, which fallback systems, which public explanations, and which emergency contracts would matter if several dependencies failed faster than the ordinary administrative tempo allows. It reveals whether the state has designed for overlap or merely for repetition.

A named example from the pandemic period clarifies this without reducing climate risk to epidemiology. During early COVID-19 response, many institutions discovered that their plans had been built less around true scenario diversity than around a narrow forecast habit. They had numbers, trend lines, and dashboards. What they often lacked was serious rehearsal of worlds in which demand surged unevenly, supply chains broke simultaneously, staff absence compounded system strain, and public communication became an operational variable rather than a final explanatory step. The lesson is not that climate planning should mimic pandemic planning in all respects. It is that institutions often discover too late that their forecasting apparatus was more mature than their capacity to think through divergent planning worlds.

A second example lies in national security and civil contingency planning more broadly, where scenario methods have long been used not to predict a single attack, crisis, or outage, but to reveal which assumptions a defence posture, emergency response plan, or continuity system has quietly standardised. This matters because it shows that scenario work is not a literary indulgence imported into serious policy from the outside. It is one of the few ways institutions can expose the structure of their own confidence. When used well, a scenario is like a pressure test for thought. It does not tell you which pipe will burst on which day. It tells you where the system is under strain before water is already on the floor.

Still, caution is necessary. Scenario work can go wrong in several familiar ways. Some institutions design scenarios so comfortable that they merely restate the present in slightly darker colours. Others design them so sensationally that the exercise becomes a contest in shock value rather than a discipline of planning. Some fill scenarios with implausible detail that flatters the authors' ingenuity while distracting from the structural trade-offs the exercise was supposed to reveal. Others generate multiple futures and then quietly continue planning against the one the institution already preferred. These are not arguments against scenario work. They are reminders that scenarios must be judged by what they surface, not by how vivid or frightening they sound.

This is where plausibility discipline begins to matter, even before the next section takes it up more directly. A useful scenario rests on mechanisms rather than headlines. It asks what could change the speed, sequence, or interaction of risks in ways that would matter to the system under review. It resists the temptation to substitute dramatic imagery for governance relevance. A scenario is well designed when it makes planners say, "if that world began to take shape, our current plan would fail here, here, and here". It is badly designed when the most memorable feature is its spectacle rather than the failure point it exposes. The purpose is not to impress. It is to sharpen judgement.

The reader should also notice that scenarios can perform a political service beyond the immediate planning room. They make disagreement more disciplined. Instead of arguing endlessly over which future is "right", institutions can compare how different choices perform across different plausible worlds. This shifts the argument from prophecy to robustness. A minister, budget officer, emergency planner, or mayor may disagree about likelihoods and still find common ground in the narrower question of which capacities remain valuable under several conditions. That is often a more productive basis for action than the pursuit of a single authorised forecast that everyone privately knows may be overtaken by events.

What scenarios finally do, when they are used seriously, is change the quality of institutional honesty. They force a plan to reveal what kind of future it secretly needs in order to succeed. They expose whether a budget assumes attrition will remain slow, whether a warning system assumes trust will remain stable, whether logistics assume mutual aid will be there, whether communications assume revision can occur without reputational cost. Once those assumptions are visible, decision-makers can begin to choose with greater candour. Not certainty. Candour.

The map had already changed in October 1962, and it changes whenever institutions are asked to act before the future has settled into one recognisable shape. The harder question is not which scenario will come true. It is whether the plans already on the table are quietly betting on one tempo, one sequence, one kind of public, and one kind of institutional composure - and what happens if the world arrives in a different order than the one those plans were written to survive.

Tail risks and plausibility discipline

In September 2008, after the collapse of Lehman Brothers, one of the most common retrospective judgments was that the warning signs had not been wholly absent; what had been absent was a planning culture willing to keep an uncomfortable chain of failure in view long enough to act on it. The institution was Lehman Brothers. The date is easy to verify. The place was New York, though the consequences travelled far beyond it. The most dangerous scenarios had often been treated as either too disruptive to contemplate seriously or too improbable to discipline ordinary decisions. The weak seam held.

That pattern is not confined to finance. It appears wherever institutions face low-probability, high-consequence outcomes that threaten the dignity of normal planning. Tail risks are difficult not only because they are frightening, but because they embarrass routine judgement. To keep them in view is to admit that the official baseline may be too comfortable, that the current budget may rest on a flattering assumption, and that the plan praised for prudence may in fact depend on a world more forgiving than the one actually unfolding. The weak seam held again, not because it was strong, but because the institution preferred not to test it too hard.

This is why plausibility discipline matters. The concept is more useful than a general appeal to "take extremes seriously", because extremes can be invoked lazily, theatrically, or opportunistically. A disciplined treatment of tail risk does not begin with the most dramatic headline available. It begins with mechanism. What chain of pressures, dependencies, delays, or failures would have to align for an extreme outcome to become administratively relevant? Which parts of that chain are already visible in weaker form? Which assumptions in the current plan would fail first if that chain began to form? The implication is exacting: a tail risk deserves attention not because it is vivid or terrifying, but because the pathway leading to it is coherent enough that ignoring it would be a form of institutional self-deception.

That distinction matters because sensationalism is one of the easiest ways to ruin serious planning. If extreme scenarios are introduced mainly as shock devices, institutions learn the wrong lesson. They either dismiss the exercise as melodrama or perform concern without changing any underlying decision rule. Plausibility discipline avoids that trap by asking for a credible sequence rather than a memorable ending. A supply system does not need to fail in the most cinematic way imaginable for tail risk analysis to be useful. It is enough to ask how repeated local disruptions, concurrent regional demands, communications degradation, and fuel constraint might move together into a wider failure that the present plan has not rehearsed. The point is not spectacle. It is stress on the decision system.

This is where institutions often become biased towards the most comfortable scenario. Comfort in planning rarely announces itself as comfort. It usually appears under more respectable names: realism, proportionality, fiscal discipline, moderation, prudence. Each can be genuine. Each can also mask a quiet preference for futures in which present arrangements remain broadly defensible. The comfortable scenario is not always the one people consciously choose because it feels nice. It is often the one that preserves existing budgets, protects established hierarchies, and allows difficult structural change to remain someone else's problem for another cycle. Tail risks become inconvenient not only because they are frightening, but because they expose how much the institution's current posture depends on a future that is administratively forgiving.

One can see this bias clearly in infrastructure and emergency planning. An agency may accept that severe disruption is possible in theory while continuing to plan around a slower, more manageable tempo because only that tempo fits the staffing model, the maintenance budget, and the mutual aid agreements currently in place. A municipality may recognise the possibility of repeated compound stress and still frame its planning around the assumption that neighbouring jurisdictions will remain available for assistance, because to do otherwise would require admitting how thin its real margin is. A ministry may acknowledge cascading failure and still fund only incremental improvements, because the more demanding scenario would imply a redesign too costly or politically awkward to face. In each case, the preferred scenario is not necessarily the most likely. It is the one the institution can live with psychologically and fiscally.

A useful example comes from pandemic preparedness before COVID-19. Many institutions possessed plans for outbreaks. They had stockpile concepts, continuity documents, emergency committees, and public communication templates. What they often lacked was a serious willingness to keep more disruptive planning worlds in view: prolonged staff absence, global supply failure, political contestation over basic public health measures, hospital strain across regions at once, and communication environments in which misinformation became an operational factor rather than a side issue. The problem was not that extreme outcomes had never been imagined. It was that the planning culture often preferred versions of the future that preserved the plausibility of normal recovery patterns. When harsher conditions arrived, the comfortable scenario was exposed less as a forecast error than as an institutional preference.

A second example lies in wildfire planning, where systems can become attached to tail-risk avoidance in the wrong form. Some agencies either over-focus on spectacular worst cases detached from operational relevance, or under-focus on them because such cases seem too rare to deserve serious restructuring. Plausibility discipline offers a better route. Instead of asking whether a fire season will become an unprecedented catastrophe in the broadest sense, one asks which mechanisms would make a severe season administratively destabilising: simultaneous regional demands, smoke affecting areas far from the fire line, power shut-offs, transport strain, mutual aid exhaustion, hospital load, and degraded public compliance. Once framed this way, the tail risk becomes useful. It no longer depends on rhetorical apocalypse. It depends on a sequence of pressures that the institution can actually test.

This points towards a practical method. A tail risk should earn planning attention when three conditions hold. The first is mechanistic coherence: the path from present conditions to extreme strain can be described in steps that are intelligible and institutionally relevant. The second is decision relevance: if the path began to materialise, current plans, budgets, or authority structures would face choices they are not prepared to make. The third is asymmetry of consequence: even if the outcome remains unlikely, the cost of having ignored it would be sufficiently high that some present adjustment in attention, rehearsal, or reversible investment is justified. None of these criteria requires certainty. All of them require discipline. They keep the analysis close to governance rather than drifting into imaginative theatre.

Still, caution is necessary. Tail-risk thinking can become a form of institutional vanity, especially in organisations that enjoy being seen as serious about catastrophe. Extreme scenarios can seduce planners into overcomplication, language inflation, and ever more elaborate maps of breakdown that no decision-maker can actually use. This is one reason the emphasis on mechanism is so important. The test of a tail-risk exercise is not whether it produces awe. It is whether it sharpens choice. Does it reveal which authority is missing, which reserve too thin, which dependency too concentrated, which communication assumption too fragile? If it does not, the scenario may be vivid and still be practically idle.

A fair counter-example should also be admitted. Institutions can overcorrect. Some become so impressed by a severe tail risk that they distort all planning around it, consuming scarce attention and resources that should also protect against more frequent and still damaging pressures. This is not discipline either. A planning system dominated by the most extreme imaginable case can become brittle in a different way: paralysed by caution, prone to false alarms, and unable to justify itself politically when the extreme does not materialise. The stronger argument is therefore not to elevate tail risks above all else. It is to keep them present enough that they can test comfortable assumptions without colonising every decision.

This balance is hard because human beings and institutions alike have a deep appetite for narrative closure. We want to know whether a severe outcome is "really" coming, whether it can be ruled out, whether it deserves full mobilisation or polite notation. Tail risks deny that comfort. They ask planners to accept a less satisfying burden: not knowing with confidence whether the extreme will arrive, while still deciding whether the route towards it is coherent enough that current plans ought to bend in response. That burden is precisely why many organisations drift back towards the comfortable scenario. Comfort offers a story in which ordinary procedures remain adequate a little longer.

The phrase "keeping tail risks in view without paralysing action" therefore points to a deeper virtue: proportionate unease. A serious institution should feel enough unease to test itself against harsh but mechanistically plausible futures, and enough discipline to keep those futures tied to choices rather than to dread. That means treating tail risks neither as fringe curiosities nor as governing destiny. It means asking what would fail first, what could be made more reversible, what trigger would matter, what reserve would buy time, what communication would need to change, and which seemingly minor weakness now could become decisive under load. This is not glamorous work. It is often the opposite. It strips drama away from extremes and translates them into administrative consequence.

The weak seam held in 2008 until it did not, and that is the image worth keeping. Tail risks are rarely useful because they predict an exact disaster. They are useful because they reveal where the seam already exists in present arrangements. The unresolved question is not whether institutions can imagine severe outcomes; they can. It is whether they can discipline themselves to look at the seam without either flinching into comfort or freezing into catastrophe, and then act just enough that the future they most fear is no longer also the one their current assumptions quietly prepare.

Red teaming and institutional stress tests

In August 2005, Hurricane Katrina exposed a truth that emergency institutions often resist until events force the matter: a plan can look coherent on paper and still fail almost immediately once authority, logistics, and communication are tested under real load. The place was the Gulf Coast of the United States. The date is easy to verify. The most famous images showed floodwater, helicopters, stranded residents, and administrative confusion. The deeper lesson was less visual and more unsettling. Systems failed not only because the hazard was severe, but because assumptions about coordination, readiness, and fallback capacity had not been challenged hard enough before the storm arrived. The table shook.

That image of the table matters because red teaming is often misunderstood as a theatrical exercise in criticism, a ritual in which clever sceptics tear holes in official plans for the sake of looking realistic. Done badly, it can become exactly that. Done well, it serves a more serious purpose. It tests whether a plan still functions when the most flattering assumptions are removed. It asks who decides when the expected decision-maker is unavailable, what happens when supplies do not arrive on schedule, whether communications remain clear when messages must change quickly, and whether neighbouring support remains available when neighbours are also in trouble. The table shook again, and the point was not to humiliate the plan. It was to discover whether the system beneath it could bear real weight.

This is the central argument. Structured challenge processes matter because institutions are naturally drawn towards narratives in which their own arrangements remain broadly workable. They assume the call will connect, the mutual aid request will be honoured, the backup generator will have fuel, the public will interpret revised guidance charitably, and the legal authority to act will be obvious when the moment comes. None of these assumptions is absurd. Many are reasonable under ordinary conditions. The concept to name here is adversarial rehearsal. It refers to the deliberate testing of plans against inconvenient but plausible realities, not in order to predict disaster exactly, but in order to reveal where confidence depends on conditions the institution has not earned. The implication is exacting: a system that never submits its own plans to disciplined internal challenge is likely to discover its weak points in public.

This is why stress testing governance is different from stress testing hardware. It is relatively easy to think in terms of physical tolerance: how much water a barrier can bear, how long a battery lasts, how far a rescue aircraft can reach, how much stock a warehouse can hold. Those questions matter. Yet public systems often fail first through governance. Decision rights may be too unclear to trigger action at speed. Logistics may depend on contracts, routes, or labour assumptions that no one has tested under regional strain. Communications may rest on the fantasy that one authoritative message will remain stable, comprehensible, and trusted even after circumstances shift. A governance stress test asks what happens when the problem is not that the infrastructure breaks immediately, but that the institutions charged with using it do not know how to move fast enough or together enough when the pattern of demand changes.

The difference is not academic. A health system can possess beds and still fail because oxygen supply, transport coordination, staffing authority, and public messaging do not hold together under concurrent demand. A municipal evacuation plan can possess routes and shelters and still fail because neighbouring jurisdictions are following similar plans at the same time, because fuel distribution narrows unexpectedly, or because the threshold for changing guidance is politically contested. A national continuity system can have contingencies for one region's failure and still collapse when several regions draw on the same reserve of contractors, airlift, or emergency management personnel. Stress testing governance therefore means asking which assumptions about availability, sequence, and cooperation remain invisible until several pressures arrive together.

This is where mutual aid deserves closer scrutiny. Mutual aid agreements are among the most comforting instruments in emergency planning because they convert scarcity into relationship. They promise that if one jurisdiction is overwhelmed, others will help. Such promises are often sincere and often useful. They are also highly vulnerable to concurrency. A simultaneous regional emergency can overwhelm mutual aid assumptions with brutal speed because the very partners counted on as fallback are now busy defending their own networks, hospitals, crews, or shelters. The agreement still exists. The capacity behind it has changed. A red team should therefore ask not only whether the mutual aid mechanism is formally sound, but whether it survives a world in which the stress is shared rather than localised. The neighbour may be busy.

A useful example can be drawn from wildfire seasons in western North America, where jurisdictions have at times faced overlapping fire conditions that strained interagency resource sharing and mutual assistance. The point is not to litigate one season or one command structure. It is to show how quickly a plan built around imported crews, aircraft, or logistical support becomes thinner when the surrounding region is drawing on the same pool. In such settings, a red team does not need to predict a singular apocalyptic event. It needs only to ask what happens if the wider system cannot spare what the plan quietly assumes it will spare. That question alone often reveals more about real preparedness than pages of baseline planning language.

A second example lies in pandemic-era stress on hospitals and supply systems, where mutual aid, shared procurement, and regional support assumptions were repeatedly tested by simultaneous demand. Again, the hazard differed from the climate-linked risks this book addresses directly, but the institutional lesson is the same. Systems designed to share burden across geography can fail when geography ceases to diversify burden. A mutual aid agreement that works under local flood or isolated infrastructure failure may prove nearly useless when heat, smoke, blackout, or storm strain several adjacent jurisdictions at once. Red teaming is valuable precisely because it forces planners to ask whether their resilience model is genuinely distributed or merely dependent on someone else's spare capacity.

Still, caution is necessary. Red teaming can fail in its own distinct ways. Some organisations turn it into staged scepticism, inviting challenge but only within safe limits that leave senior assumptions untouched. Others fetishise disruption, constructing ever more punishing scenarios that reveal everything and therefore prioritise nothing. Some mistake contrarianism for rigour, as though the most valuable challenge were always the one most opposed to the current plan. None of that is helpful. Structured challenge works only when it remains tethered to plausible mechanisms, specific decision points, and the actual authorities, contracts, communications channels, and reserve assumptions on which the institution depends. The aim is not to prove that every plan fails. The aim is to identify *how* it fails, *where* it fails first, and *which* adjustments would materially widen the room for competent response.

This is why communications deserve inclusion as a core test, not a closing supplement. Institutions often red team logistics and command chains more readily than public voice. That is a mistake. A revised evacuation boundary, a shelter closure, a fuel advisory, or a change in medical guidance can all become destabilising if the communication architecture assumes stable facts, abundant trust, or singular authority. A useful stress test therefore asks what happens when messages must change midstream, when local and national voices do not align perfectly, when rumours outpace formal correction, or when revised guidance is accurate but politically awkward. These are not peripheral concerns. They are part of whether the logistics plan remains usable once people respond to it.

The political value of red teaming lies in what it does to institutional culture. Most public systems quietly reward confidence, procedural smoothness, and the ability to present plans as settled. Structured challenge interrupts that reward pattern. It makes it professionally legitimate to ask whether the fallback authority is real, whether the reserve stock is actually reachable, whether the private contractor would still answer during shared regional disruption, and whether the communications plan presumes trust not yet earned. This can feel abrasive. In truth, it is a form of respect. It treats the institution as capable of looking at its own illusions before the public has to live with them.

A fair counter-example should be admitted. Some plans perform reasonably well without elaborate red teaming because the hazard domain is narrow, the operational chain short, and the authorities comparatively centralised. Not every institution needs a permanent adversarial machinery. In some contexts, straightforward exercises, after-action review, and well-run technical audits may surface the relevant weaknesses. That matters because the case for structured challenge should not become a new orthodoxy of permanent internal warfare. The deeper point is that climate feedback risks, compound emergencies, and degraded mutual aid conditions increase the chance that comforting assumptions will survive ordinary review. Where that chance rises, challenge has to become more organised.

The measure of success is not whether an exercise produces embarrassment. It is whether it changes subsequent decisions. Did authority lines become clearer? Did reserve assumptions change? Did mutual aid get reclassified from guarantee to contingent support? Did procurement or staffing plans reflect the possibility of simultaneous regional need? Did communications triggers become more explicit? A red team that ends in a polished report and no altered behaviour is merely a more sophisticated way of admiring the problem. The useful version leaves a mark on how the system allocates attention, money, discretion, and public explanation.

The table shook in 2005, and it will shake again wherever institutions imagine that plans are tested by being written rather than by being pressured. The harder question is not whether organisations can tolerate criticism in principle. It is whether they can build challenge processes sharp enough to expose where decision rights blur, where logistics depend on unavailable generosity, and where communications assume a calmer public than any real crisis is likely to permit - before the next simultaneous emergency makes those assumptions visible in the only way left.

Trigger points and adaptive pathways

In August 2017, Hurricane Harvey struck Texas and forced public authorities to make decisions under conditions that were both urgent and unstable. The place was Texas. The date is easy to verify. Rainfall totals, flood extent, infrastructure failure, evacuation choices, shelter demand, and public messaging all evolved faster than many formal plans were designed to absorb. What mattered was not only what officials knew at any given hour. It was whether they had already decided what certain kinds of warning, threshold, or system strain would *mean* in practice. The switch had to move.

That image is useful because institutions often imagine that monitoring and action naturally belong together. They do not. A river gauge can rise, a heat alert can intensify, permafrost deformation can worsen, fuel reserves can fall, or outage duration can lengthen without any of these signals automatically producing decision. Data tell a system that conditions are changing. They do not by themselves decide who should act, when discretionary authority becomes mandatory, which budget line can be opened, whether public communication should shift tone, or whether the current plan has crossed from strain into failure. The switch had to move again, and in many institutions that is precisely where uncertainty becomes politically dangerous: not in the measurement, but in the absence of a pre-agreed pathway from measurement to action.

This is the central argument. Trigger points should be treated as governance commitments, not merely as technical thresholds. A threshold says that something has changed in the world. A trigger says that the institution has already committed itself to treating that change as a reason to do something specific. The distinction is small in wording and enormous in practice. The concept to name here is conditional commitment. It refers to the decision, made before crisis fully matures, that when a defined condition is met, certain actions, reviews, escalations, or public disclosures will follow unless a named authority decides otherwise and records why. The implication is exacting: without such commitments, monitoring can become a theatre of attention in which officials know more and do little until the costs of delay become public.

This is why adaptive pathways matter. They acknowledge that institutions cannot know the whole future in advance, yet they refuse the opposite error of waiting passively for certainty. An adaptive pathway links present choices to future possibilities through a chain of monitored conditions and pre-agreed responses. It does not demand clairvoyance. It demands preparedness about process. A road authority may not know exactly when a route will become too unreliable for current maintenance assumptions. A health system may not know precisely when heat will begin to overwhelm ordinary staffing patterns. A coastal municipality may not know when recurrent damage will cross from tolerable inconvenience into untenable fiscal burden. What it can know, or at least decide, is what evidence would trigger deeper inspection, temporary restrictions, procurement acceleration, emergency communication, design review, or strategic retreat from older assumptions.

The practical value lies in moving argument upstream. Without triggers, each worsening condition becomes a fresh political contest. Officials debate whether the warning is serious enough, whether acting now would look excessive, whether waiting one more cycle might preserve fiscal calm, whether the public can tolerate revision, whether another office should move first. With a trigger-based pathway, some of that contest has already been disciplined. The argument is not abolished. It is bounded. The institution has said in advance: when this signal appears in this form, we will review, escalate, spend, disclose, suspend, or redesign according to a known pathway. This does not eliminate uncertainty. It makes uncertainty governable.

A named example can be found in Dutch water management, where much of the practical seriousness of adaptation has long depended not only on technical knowledge but on institutional pathways linking observation, review, and action. One need not romanticise the Netherlands or claim that its arrangements solve every problem to see the broader lesson. Where thresholds are embedded in a wider culture of responsibility, maintenance, and revision, monitoring becomes more than passive awareness. It becomes part of a public promise that changing conditions will not simply be watched. The relevance here is not that every country can replicate Dutch institutions. It is that trigger points only gain meaning when they are nested inside governance systems that know who decides and who pays when a threshold is crossed.

A second example can be drawn from drought and reservoir management, where authorities often possess extensive monitoring data yet still struggle over when exactly to move from conservation appeals to restrictions, from restrictions to priority reallocations, or from technical concern to overt political communication. The difficulty is rarely measurement alone. It is commitment. If the pathway from falling levels to stronger action remains ambiguous, each step can be delayed by fear of overreaction, economic pressure, or concern about public backlash. The result is that the threshold may be visible long before it is institutional. A trigger becomes useful when it marks not just the state of the reservoir, but the state's own prior decision about how it will govern that state.

This is why reversibility deserves such emphasis. Institutions often fear triggers because they imagine that every triggered action locks them into a final, politically costly course. Sometimes it does. Often it need not. A mature adaptive pathway distinguishes between actions that should be reversible and actions that should not. Early restrictions, temporary staffing shifts, precautionary communication changes, additional inspections, conditional procurement, and pilot infrastructure measures can all be designed as reversible if conditions improve or if later information changes the picture. Reversibility matters because it lowers the political and administrative threshold for acting in time. Officials are more likely to move before the crisis is undeniable if they know that not every triggered step is a one-way declaration of failure.

That point matters especially in public communication. One reason institutions hesitate to act on early triggers is that they fear being accused of alarmism if the worst outcome does not arrive. This is a real risk. It can be reduced when pathways are explained as staged responses rather than as all-or-nothing judgments. If authorities can say, in effect, that a trigger has been crossed and therefore a defined precautionary phase has begun, they are framing action as part of an accountable process rather than a panicked leap. That framing helps preserve trust because it tells the public that movement does not mean chaos. It means the institution is doing what it said it would do when certain conditions appeared.

Still, caution is necessary. Trigger systems can fail in both directions. Some are too vague to compel anything. They look impressive in policy documents but leave enough room for political evasion that crossing the threshold changes nothing operationally. Others are too rigid. They treat complex environments as if one variable or one threshold can legitimately govern a whole chain of consequences. Such systems may generate automatic responses that fit the metric more than the reality. They can also encourage gaming, in which actors manage the appearance of the indicator rather than the underlying risk. A serious system therefore needs triggers that are specific enough to matter, but nested enough in professional judgment and review that they do not become mechanical substitutes for governance.

This is why learning has to be built into the pathway itself. A trigger is not a sacred line discovered once and for all. It is a hypothesis about when institutional action should change. After each use, or each failure to use it, the institution should ask whether the threshold was too late, too early, too narrow, too broad, too politically exposed, or too disconnected from the actual burden on services and communities. Learning of this kind is not glamorous. It is often resisted because it implies that a carefully negotiated threshold was never final. But the alternative is worse. A trigger that cannot be revised becomes either ritual or trap. Adaptive planning is credible only when its own instruments are subject to adaptation.

The governance burden becomes sharper when several triggers interact. A heat threshold may coincide with power stress. A rainfall threshold may coincide with road fragility and care-home risk. A coastal erosion signal may intersect with insurance withdrawal, school transport strain, and political pressure over relocation. Institutions often create triggers one hazard at a time because agencies are organised that way. Real disruption is less polite. A good pathway therefore does not only specify what happens when one metric crosses a line. It also asks who has authority when several thresholds are breached across sectors at once and ordinary departmental sequencing breaks down. This returns us to the earlier concerns of the book: decision rights, logistics, communications, and trust.

A fair counter-example lies in systems that work tolerably well without elaborate trigger architecture because local knowledge, administrative discretion, and repeated experience are already strong enough that officials can move informally with reasonable confidence. Such systems exist. They should not be dismissed. Yet they often depend on people, habits, and tacit knowledge that may not survive turnover, scale, or compounding stress. Formal trigger-pathway design becomes more valuable as systems grow more complex, politically contested, and exposed to multiple pressures at once. It preserves memory. It makes commitments visible. It helps a government act as a system rather than as a loose collection of professionals hoping that judgment will coincide at the right moment.

This is why trigger points provide a practical bridge from uncertainty to accountable planning. They do not promise that the right decision will always be made at the right time. They do something harder and more realistic. They require institutions to state, before the full crisis arrives, what kinds of signal will count as reasons to move, who is authorised to move, what form that movement will take, and how the action can later be reviewed and revised. That is what turns monitoring into governance.

Without it, uncertainty remains a reason to wait. With it, uncertainty becomes a reason to prepare for staged action without pretending the future is fully known.

The switch had to move in Texas, as it has to move in every system that claims to monitor rising risk while remaining institutionally free not to act on what it sees. The unresolved question is not whether governments can build better sensors, sharper models, or more detailed dashboards. It is whether they can bind themselves, credibly and publicly enough, to pathways that make those signals matter before the next threshold is crossed and everyone discovers that the warning was real but the commitment was not.

Chapter 13

Critical Infrastructure and Readiness Metrics

What becomes critical when the system is already under strain? The question sounds administrative, almost dry, until one notices how quickly the answer changes once disruption spreads across several services at once. A power outage is rarely only a power outage. A damaged road is rarely only a transport problem. A communications failure can move silently into clinics, pharmacies, dispatch systems, and home care before any official list has time to catch up. The label comes late.

That delay matters because states and organisations often speak of critical infrastructure as though it were a stable category waiting to be discovered by experts and then protected accordingly. In practice, criticality is partly technical and partly political. It reflects what a society fears losing, what it notices first when it fails, and which dependencies it has learned to treat as background rather than as active conditions of survival. Lists can therefore become flattering documents. They preserve a picture of priority that feels orderly on paper while the real system beneath them has already changed shape. The label comes late again, and the costs of lateness are carried in ordinary places: repair queues, missed appointments, stalled deliveries, delayed restoration, and public surprise at how much depended on what had barely been mentioned.

This is the chapter's governing tension. Modern systems are organised around interdependence and optimisation, while readiness depends on recognising that interdependence can rearrange what counts as indispensable. A communications network that once seemed secondary to healthcare infrastructure may become central once records, consultations, pharmacy coordination, dispatch, and household care all depend on it. A warehouse may matter more than a headline policy. A maintenance contractor may matter more than a national strategy during the first difficult days of restoration. Critical infrastructure, then, is not simply a matter of named assets. It is a matter of which functions hold the rest of the system together when strain starts to travel.

That is why the stakes are larger than classification alone. To define something as critical is to direct money, attention, contingency planning, and political seriousness towards it. To omit something is not merely to make a technical mistake. It is to accept, often silently, that failure in that domain can be absorbed elsewhere. Sometimes that judgment is defensible. Often it is inherited rather than tested. Climate stress makes this inheritance more dangerous because dependencies do not remain fixed. They deepen, migrate, and sometimes invert. A system that once survived with thin inventories, lean staffing, and just-in-time logistics may still look efficient while quietly losing its capacity to absorb delay. Readiness is then weakened not through one dramatic cut, but through a long series of acceptable adjustments that no single office would describe as a strategic gamble.

This is also why resilience can disappear invisibly. It is easier to celebrate the removal of waste than to notice the disappearance of slack. Spare parts look idle until they are needed. Extra storage looks inefficient until a route closes. Redundant communications channels seem excessive until the main one fails under load. Reserve staff appear expensive until several teams are absent at once. The political bias towards visible output and neat accounting makes this especially hard to correct. Budgets reward what can be counted easily. Readiness often depends on what looks underused in calm periods. The result is a dangerous asymmetry: systems can be hollowed out through incremental optimisation long before anyone admits that what has been removed was not waste but shock tolerance.

Logistics sharpens this asymmetry further because it remains both indispensable and curiously easy to ignore. High-level plans often assume movement without dwelling on the means of movement. Supplies will arrive, crews will deploy, repairs will begin, mutual aid will cross the region, fuel will be available, warehousing will absorb the surge. Yet logistical systems are where plans meet geography, labour, maintenance, and time. A narrow road, a damaged bridge, a full depot, a missing part, an unavailable crane, a shortfall in lift capacity, or an exhausted repair crew can turn a funded response into an inert promise. This is why logistics so often behaves like hidden sovereignty inside a crisis. It determines which intentions remain real after the press conference ends.

The chapter's movement follows that reality from definition to consequence. It begins by asking what societies mean when they call infrastructure critical, and how that judgment shifts once dependencies under stress become visible rather than assumed. It then turns to the quieter capacities - redundancy, buffers, spare stocks, staffing depth, storage, and strategic slack - through which systems remain governable under interruption, before moving into logistics as the practical backbone that decides whether response plans can travel from paper into action. From there the focus settles on measurement itself, where seemingly neutral indicators can either sharpen readiness or reward the appearance of control while leaving failure modes untouched.

What unifies these concerns is the politics of attention. Measurement is not a passive mirror held up to reality. It teaches institutions what to watch, what to fund, what to report upward, and what to explain away. A metric that rewards completed paperwork, nominal compliance, or average performance under ordinary conditions may flatter a system that is brittle under compound strain. A better metric asks harder questions: what assumptions does this measure hide, what forms of overlap does it miss, what decision rights does it connect to, and what exactly would change if the indicator worsened tomorrow? Once measurement is understood in this way, infrastructure readiness stops looking like a purely technical field. It becomes a struggle over what kind of evidence deserves the power to move budgets and priorities.

This is why the final movement of the chapter matters beyond its own immediate topic. Institutions reform themselves through what they choose to count and what they allow to remain background noise. If communications continuity is measured only as uptime, not as its role in preserving care, dispatch, and public trust under compound disruption, then the wrong lessons will travel into the next budget. If logistics is treated as a support function rather than as a determinant of whether strategic commitments can be honoured, then restoration promises will keep outrunning actual capacity. The same is true of buffers, warehousing, and staffing. What is omitted from readiness metrics is often what later reappears as surprise.

The label comes late unless the system learns to ask earlier and more honestly what really holds during strain, what only appears to hold because the weather has been kind, and what has already been optimised past the point where it can still count as preparedness. That is the friction at the heart of this chapter: whether the infrastructures a society calls critical are truly the ones that preserve continuity under stress, or merely the ones it has grown used to noticing after failure has already travelled through everything else.

What counts as critical, and why

In March 2020, as hospitals in many countries prepared for surging demand during the early phase of the pandemic, one lesson emerged with unusual speed: healthcare continuity depended on systems that many public lists did not habitually place at the emotional centre of medical care. The date is easy to verify. The places varied, but the pattern was strikingly consistent. Beds, clinicians, and oxygen mattered, of course. So did records, communications links, dispatch systems, remote consultation platforms, pharmacy coordination, device connectivity, and the ability of households to remain in contact with care providers when physical movement became harder or riskier. The list had shifted.

That shift is the right place to begin because "critical infrastructure" is rarely a natural category waiting to be discovered. It is a judgment. Societies decide, explicitly or implicitly, which systems count as indispensable, which can be restored later, which failures are politically tolerable, and which dependencies are left in the background because they do not fit older imaginations of what public life rests upon. Energy, water, transport, healthcare, and finance usually appear early in such judgments, and for good reason. Yet criticality is never only about the size or visibility of a sector. It is about what happens elsewhere when that sector falters. The list had shifted again, not because hospitals had become less important, but because the path by which hospitals remained functional had become harder to ignore.

This is the central argument. Critical infrastructure should be understood not as a fixed list of prestigious assets, but as a changing map of interdependence. The concept to name here is relational criticality. A system becomes critical not solely because it is large, expensive, or symbolically important, but because other essential functions begin to fail quickly when it is lost. The implication is exacting: what counts as critical reveals a society's priorities, but it also reveals how well that society understands its own dependencies. If the formal list no longer matches the actual pathways through which continuity is sustained, the state is governing an older version of the system than the one people now inhabit.

This matters because formal critical infrastructure designations often carry real consequence. They shape protection priorities, investment patterns, staffing attention, reporting obligations, contingency planning, procurement rules, and public narrative. Once a sector or asset is placed inside the circle of the critical, it attracts seriousness. Once it remains outside, it may still matter enormously in practice while receiving less institutional focus than the interdependent system requires. Classification is therefore not a technical housekeeping exercise. It is a political act. To call something critical is to say, in effect, that its failure would expose unacceptable vulnerability in the wider social order. To omit something is to claim, often silently, that failure there can be absorbed elsewhere.

The difficulty is that climate stress changes the meaning of absorption. Systems that once failed locally and recoverably can become more consequential when disruption overlaps, when repair intervals shorten, when backup capacity thins, or when service delivery becomes more digitally and logistically dependent than older planning categories assumed. Criticality can therefore migrate. A communications network once treated as a support service may become central to healthcare continuity when electronic records, telemedicine, ambulance dispatch, pharmacy confirmation, family notification, specialist consultation, and home monitoring all depend upon it. A warehouse may become more operationally significant than a visible headquarters. A transport link may matter less for economic throughput in the abstract than for the restoration of electricity, water, and clinical capacity after a regional disruption. The system changes first. The classification often changes last.

This lag between lived dependence and official recognition is one of the most important readiness problems in modern governance. Institutions prefer stable categories. Stability simplifies budgeting, regulation, and oversight. It also flatters existing bureaucratic maps. Yet interdependence evolves faster than institutional naming practices do. Health services provide the clearest example. They are rarely self-contained in the way public imagination suggests. Hospitals depend on electricity, water, transport, medical gases, staffing mobility, data systems, cleaning supply, pharmaceuticals, waste removal, and communications. Primary care and home care are even more dependent on connective infrastructure than their modest physical footprint suggests. Under compound strain, communications can move from useful adjunct to decisive enabler. A clinic may still stand, the staff may still be willing, the medicines may still exist somewhere in the system, and yet continuity can still collapse if records cannot be accessed, referrals cannot be confirmed, dispatch cannot coordinate, or patients cannot be reached.

That is why the example of communications infrastructure becoming critical to healthcare continuity is more than a contemporary curiosity. It reveals the weakness of sectoral thinking. Public lists often classify communications and healthcare as separate domains, each important in its own right. Useful enough, but incomplete. What matters under stress is not only whether each domain survives independently. It is whether the relationship between them has become so tight that losing one function effectively disables the other. Once that is true, criticality belongs not only to the sectors but to the connection between them. A state that protects the hospital building while neglecting the data links, dispatch continuity, and communications redundancy on which modern care depends is protecting the symbol while underprotecting the service.

A named example from Hurricane Maria in Puerto Rico is instructive here. Much of the public focus fell, understandably, on the electricity crisis after the storm. Yet the practical breakdown of healthcare continuity also depended on communications failure, disrupted records access, coordination difficulties, interrupted supply chains, and the inability to connect patients, providers, pharmacies, and emergency transport reliably across a damaged system. The lesson is not that communications became more important than power. It is that criticality under stress was distributed across an interacting set of dependencies, some of which official discourse and planning traditions had treated more separately than real continuity allowed. Once the system was pressured, the interconnection became brutally visible.

A second example can be drawn from remote and rural healthcare systems, where communications links have become increasingly central to continuity even outside formal emergency. Telehealth, digital triage, specialist consultation, diagnostics transmission, prescription management, and coordination between community care and hospitals all rely on network continuity that earlier models of healthcare criticality might have treated as peripheral. Under climate-linked disruption, these dependencies grow sharper. Extreme weather or infrastructure damage can constrain travel at the same moment that patients depend more heavily on digital contact to avoid unsafe movement or to preserve continuity while roads, ferries, or air links are compromised. The critical service is no longer only the clinic in place. It is the clinic in relation.

This is why readers should question whether official critical infrastructure lists match real system interconnections. The question is not whether lists are useless. They are necessary. The sharper question is what assumptions the list hides. Does it privilege large, visible, and politically salient assets over quieter dependencies? Does it classify sectors separately where continuity actually depends on their overlap? Does it reflect historical strategic anxieties more than present operational vulnerability? Does it assume that backup systems remain available, localised failure remains localised, and digital mediation remains optional rather than structural? A good list can help coordinate seriousness. A stale one can preserve the appearance of seriousness while misdirecting it.

Still, caution is necessary. Criticality cannot expand without limit. If everything becomes critical, the category collapses into moral earnestness and loses operational value. States must prioritise. They must distinguish between systems whose failure would rapidly degrade wider continuity and those whose loss, while painful, remains more containable. This is one reason the concept is politically difficult. Every sector can make a plausible case for its importance. A serious approach therefore resists inflation. It does not say that all dependencies are equally decisive. It asks instead which dependencies transform local failure into system failure, which ones accelerate compound stress, and which ones are difficult to substitute quickly once lost.

A fair counter-example should also be admitted. Some formal critical infrastructure frameworks do evolve thoughtfully and already include communications, digital services, and cross-sector dependencies more explicitly than older approaches did. That matters because the problem is not universal administrative blindness. The stronger claim is that institutional understanding often develops unevenly, and that the speed of interdependence can outrun the speed of classification. Some systems have corrected for this. Others still rely on older intuitions about what "real" infrastructure looks like. The point is not to sneer at lists from the comfort of hindsight. It is to treat them as hypotheses that require repeated challenge rather than as settled truths.

This challenge is especially important under climate stress because criticality can shift not only through technology but through tempo. A communications network may appear resilient under isolated outages and become decisive under prolonged heat, smoke, flood, or repeated storm interruption because the burden on health coordination changes. A road may appear ordinary until repair crews, fuel transport, and medical evacuation all depend on it in the same difficult week. A warehouse may seem secondary until spare parts, water treatment chemicals, and backup equipment must all move through a constricted supply chain. Climate stress does not merely threaten infrastructure. It can reorder the hierarchy of what matters most.

The deeper point, then, is that what counts as critical reveals more than technical risk. It reveals what a polity has chosen to notice about itself. Which continuities it values. Which failures it is prepared to treat as intolerable. Which dependencies it still imagines as optional. The critical list is a mirror, though often an outdated one. The unresolved question is whether institutions are willing to keep asking, under changing conditions, whether the mirror still reflects the real system - or whether they will continue to protect the things they have long called critical while continuity quietly migrates elsewhere.

Redundancy, buffers, and strategic slack

In March 2021, the container ship *Ever Given* ran aground in the Suez Canal and briefly interrupted one of the world's most important trade routes. The place was Egypt. The date is easy to verify. The spectacle drew attention because it was visible, almost absurdly so: one vessel lodged across a narrow passage. Yet the deeper lesson lay elsewhere. A system praised for efficiency had left itself exposed to interruption at a point where delay multiplied quickly through inventories, schedules, repair cycles, and contracts far from the canal itself. The shelf looked full.

That image is useful because resilience is often lost not in dramatic cuts but in a long sequence of reasonable decisions that make systems leaner, faster, and cheaper under normal conditions. Warehouses shrink. Spare parts are held centrally rather than locally. Teams are reduced to what current demand seems to justify. Supplier lists are narrowed in the name of efficiency. Backup systems are retained only where regulation compels them. No single change appears reckless. Each can be defended with neat language about productivity, discipline, and modernisation. The shelf looked full again, because the systems surrounding it had been arranged to make absence visible only when the route closed or the demand surged.

This is the central tension. Efficiency and resilience are not simple opposites, but they do pull in different directions. Efficiency seeks to reduce idle capacity, duplication, and apparent waste. Resilience depends, at least in part, on preserving forms of underused capacity that only reveal their value during disruption. The concept to name here is strategic slack. It refers to the storage, staffing depth, spare parts, fallback suppliers, maintenance margin, and procedural room that allow a system to absorb interruption without immediate collapse or desperate improvisation. The implication is exacting: what looks like excess in calm periods may be the very capacity that keeps continuity credible when several pressures arrive together.

This is why redundancy should not be understood too narrowly. Public debate often treats redundancy as a crude matter of one backup system shadowing another identical one. That is part of the story, but only part. Real redundancy can take several forms. It may appear as distributed storage rather than a single large depot. It may mean technicians with overlapping skills rather than one officially designated specialist. It may mean a second communications path, a substitute supplier, an extra transformer, an additional warehouse, a reserve crew, or a maintenance schedule designed with room for delay. What unites these arrangements is not duplication for its own sake. It is optionality under strain. They create more than one way for the system to remain functional when the preferred path is obstructed.

Buffers perform a similar role but with a different temporal logic. A buffer buys time. Fuel stocks, medicine inventories, parts on hand, cash reserves, staff float, and surge transport capacity do not solve disruption by themselves. They slow the speed at which disruption becomes systemic. That slowing matters enormously. Time is often the one thing institutions cannot manufacture once a crisis is already under way. A buffer gives managers room to verify information, communicate honestly, shift contracts, reroute crews, or escalate authority before the public experiences total failure. When buffers are absent, every problem becomes immediate and political at once. The system loses not only margin but composure.

This is why strategic slack should be judged by what it enables, not by how idle it appears. A spare part on the shelf can look wasteful in a quarterly review. It looks different when a critical pump fails and replacement lead times stretch under regional demand. An apparently overstaffed maintenance function can seem indulgent until storm damage or heat stress generates a workload that cannot wait for the next tender cycle. A reserve warehouse can look inefficient until one transport corridor is interrupted and the central depot becomes too distant to matter. Slack does not justify itself through ordinary throughput. It justifies itself through continuity under abnormal load.

A clear example emerged during the early years of the COVID-19 pandemic, when many health systems and governments discovered that stockpile logic, staffing depth, and supplier diversity had been narrowed too aggressively for the system to absorb simultaneous stress. The lesson is not confined to public health. It concerns governance under disruption. Institutions that had optimised procurement and inventory around normal demand found that normal demand was an inadequate teacher. Buffers once dismissed as excessive suddenly appeared prudent. The issue was not simply shortage. It was the speed with which a lack of slack converted manageable pressure into public emergency. In those moments, the shelf no longer looked full. It looked like a memory of a different operating philosophy.

A second example lies in electricity and utility restoration after severe weather. Publics often assume that if a system is funded and technically sophisticated, repairs will follow quickly once the damage is known. In practice restoration can depend on mundane forms of slack: the number of trained line crews, the availability of poles and transformers, the existence of local storage, the ability to borrow crews from neighbouring regions, and the transport capacity needed to move equipment where it is needed. A utility may have an excellent control room and still struggle because the spare parts are too far away, the qualified workforce too thin, or the mutual aid pool too stretched by concurrent events. Here again, resilience does not fail through one dramatic decision. It erodes through incremental optimisation until the first truly difficult week reveals what has been shaved away.

This is why the chapter avoids prescribing fixed levels of redundancy or buffer. No universal number would survive serious scrutiny across sectors, geographies, or political economies. A hospital, a water system, a telecoms network, and a remote fuel depot do not need the same kind of slack, nor do they face the same trade-offs. What matters instead are evaluation questions sharp enough to reveal where resilience may be disappearing invisibly. If this route closes, how long before service quality degrades meaningfully? If this supplier fails, what alternative exists in practice rather than in contract language? If several adjacent jurisdictions are affected at once, which reserves remain genuinely local? If demand rises and staffing falls together, what work is deferred first, and what follows from that deferral? If an important part fails, where is the nearest usable replacement and who is authorised to move it? Such questions do not produce a single ideal ratio. They do produce a more honest picture of whether the system's apparent efficiency rests on fragile assumptions about calm conditions.

Still, caution is necessary. Slack can be designed badly. Stock can expire. Spare parts can become obsolete. Extra capacity can turn into administrative complacency. Warehousing can be expensive and badly located. Staffing cushions can become politically untouchable even when skill mix no longer matches need. A serious account cannot pretend that every buffer is wise simply because it is a buffer. Nor should resilience language become a moral shelter for weak management. The stronger claim is narrower and more useful. Institutions should know what kinds of slack they rely upon, what those forms of slack cost, what risk they are supposed to absorb, and what would happen if they were further reduced in the name of efficiency. Waste is real. So is hidden fragility.

A fair counter-example lies in systems that have become more efficient without becoming more brittle. Better information systems, decentralised storage designed intelligently, modular infrastructure, stronger maintenance analytics, and multi-skilled teams can all improve resilience while reducing certain forms of visible redundancy. This matters because it prevents a lazy nostalgia for bulk and overcapacity. The issue is not whether a system looks lean or heavy. It is whether it retains room to absorb interruption without immediate loss of function. Some institutions achieve this well. Many do not. The difference lies less in ideology than in whether optimisation was pursued with a disciplined understanding of dependencies, concurrency, and restoration time.

The political difficulty is that strategic slack rarely announces itself as a public good until after it has been cut. A spare warehouse, a reserve crew, a local inventory of critical components, or a conservative maintenance margin can all appear to taxpayers, auditors, and ministers as inefficiencies waiting to be corrected. The argument for retaining them is inherently harder because their value is counterfactual. They are meant to stop certain failures from becoming visible. Success therefore looks like non-event, and non-event is difficult to defend in systems hungry for measurable output. This is one reason resilience is often lost invisibly. The gains from optimisation are concrete and immediate; the losses are delayed, distributed, and only fully understood when recovery takes longer than anyone publicly expected.

This is where the relationship between efficiency and legitimacy becomes sharper. If a system has no meaningful buffers, each disruption becomes a public test of whether the state can still protect continuity. A shortfall in spare parts becomes a prolonged outage. A thin maintenance workforce becomes repeated delay. A narrow supplier base becomes visible dependence. Citizens rarely experience these as neutral trade-offs made rationally in calmer years. They experience them as broken promises about basic service. Strategic slack therefore supports not only operations but credibility. It helps preserve the gap between disruption and public breakdown, which is often where legitimacy lives or dies.

The shelf looked full when the route was open and demand was ordinary. That is the seduction of optimisation. The harder question is not whether institutions can become leaner; many can. It is whether they can still say, with a straight face and a realistic understanding of their own dependencies, that what has been removed was truly waste rather than the thin but essential margin by which a system remains governable when the ordinary timetable fails.

Logistics as the hidden determinant

In September 2017, after Hurricane Maria struck Puerto Rico, the public saw collapsed power lines, damaged roads, blocked ports, and desperate demand for restoration. The place was Puerto Rico. The date is easy to verify. Yet one of the hardest truths the disaster exposed was that plans, declarations, and funding commitments do not restore systems by themselves. Equipment has to move. Crews have to arrive. Warehouses have to release what is needed. Fuel has to reach vehicles and generators. Repairs do not occur where the decision is made. They occur where the part, the person, and the route finally meet. The road decides.

That image matters because logistics is routinely treated as support rather than substance. It appears late in speeches and low in prestige. Strategies discuss resilience, budgets announce investment, agencies issue plans, and public debate gravitates towards leadership and blame. Meanwhile the actual capacity to respond rests on less glamorous questions: where the spare parts sit, how many crews can move at once, which roads can take heavy loads, what warehousing is local rather than distant, how much lift capacity exists for urgent equipment, whether maintenance teams can reach the site before the weather closes again, whether procurement can translate authorisation into movement. The road decides again, because without logistical reality every higher-level intention remains suspended above the crisis rather than inside it.

This is the central argument. Transport, warehousing, and maintenance form the backbone of response capacity, and they do so whether institutions admit it or not. The concept to name here is logistical sovereignty. It refers to the practical ability of a system to convert command, money, and policy into timely physical effect. The implication is exacting: a government, utility, hospital network, or emergency service may possess formal authority and ample plans yet remain weak in the only sense that matters under disruption if it cannot move people, equipment, fuel, parts, and information through damaged or strained space. Response capacity is therefore never only what the institution owns or budgets. It is what the institution can *move*.

This matters because logistical constraints can quietly negate high-level plans without ever refuting them on paper. A resilience strategy may assume backup generators can be installed, but if generators are in a distant depot and heavy lift capacity is scarce, that assumption is a comfort rather than a capability. A mutual aid agreement may promise outside crews, but if airports are congested, ferries irregular, roads blocked, or housing for incoming staff unavailable, the agreement arrives in theory and not in time. A ministry may release funds rapidly after a disaster, yet funds do not clear debris, repair substations, or restock clinics unless a logistical chain already exists or can be improvised under pressure. The danger lies precisely in this gap between authorised action and physically executable action. Many institutions overestimate readiness because they measure the first and under-measure the second.

The example of limited lift capacity delaying restoration work captures this problem in concentrated form. Heavy equipment, poles, transformers, pumps, water treatment units, mobile clinics, bridge components, fuel bladders, and specialist crews often depend on constrained transport assets even in calm conditions. Under regional disruption, those constraints sharpen. A limited number of helicopters, cargo aircraft, ferries, barges, cranes, or heavy-haul trucks can suddenly become the true governor of recovery time. Public debate may continue to focus on budget release, command structure, or visible political resolve. Meanwhile restoration slows because the physical means of lifting and moving the right item to the right place are finite and already overcommitted. A system that has not mapped those bottlenecks is not only underprepared. It is planning in abstraction.

This is why single points of failure in supply routes deserve the same seriousness often reserved for critical assets themselves. A hospital may possess backup power, but if fuel reaches it through one vulnerable road or one fragile supplier arrangement, the road and the supplier become part of the hospital's real critical infrastructure. A coastal town may appear connected to the wider grid of service, yet if one port handles most heavy goods and one weather window governs resupply, continuity is only as strong as that narrow logistical channel. A utility may have equipment in regional storage, but if one warehouse, one contractor, or one bridge stands between storage and installation, then the apparent redundancy upstream collapses into concentration downstream. To ask where the system breaks is often to ask where its movement narrows.

A named example from Texas during the February 2021 winter storm reveals this clearly. Much of the public conversation centred on generation, regulation, and political accountability, rightly so. Yet beneath the visible energy crisis sat logistical realities: constrained fuel delivery, difficult road conditions, repair access, workforce strain, and the physical problem of moving equipment, supplies, and technicians through an environment suddenly hostile to ordinary transport assumptions. The lesson is not that logistics alone explains the event. It is that even when causes lie higher up in system design, restoration and mitigation are still governed by what can physically move. A frozen road or delayed fuel convoy can become as strategically important as the plant or line it serves.

A second example lies in island and remote-region restoration after storms, where limited port throughput and airlift capacity can turn every repair into a queue. Equipment arrives, but not all at once. Crews are ready, but not all where they are needed. Warehouses hold stock, but not always in the right geography or sequence. Restoration plans often assume rational prioritisation will solve the problem. In reality, prioritisation cannot conjure lift capacity from nowhere. Where heavy transport, cranes, barges, or all-weather routes are scarce, high-priority assets still wait their turn. This is one reason the logistics question is so unforgiving. It disciplines ambition with geography, weather, and timing long after the plan has been approved.

This is also why warehousing deserves more respect than it usually receives. Public systems often treat storage as passive, almost embarrassing, especially in cultures trained to praise lean inventory and centralised efficiency. Yet the location, variety, and accessibility of stock can decide whether disruption remains local or becomes systemic. A distant central warehouse can be efficient under routine conditions and practically useless under route interruption. A smaller regional stock can look expensive in peacetime and decisive when travel windows narrow. The issue is not simply quantity. It is fit between storage pattern and failure pattern. What sits on a shelf matters less than whether the shelf is reachable under the conditions that will matter most.

Maintenance belongs in the same sentence because logistics is not only about crisis movement. It is about whether the system enters crisis already weakened by deferred work and thin repair capability. A society can spend heavily on new assets and still degrade readiness if it underinvests in the supply chains, training, contractor networks, and spare parts inventories that make maintenance credible. Under climate stress, this becomes sharper. More frequent or overlapping disruption shortens the interval between incidents, which means maintenance backlogs are no longer merely an accounting concern. They alter starting conditions for the next shock. A maintenance team delayed by lack of parts today may become the reason a second, otherwise manageable failure becomes public crisis next month.

Still, caution is necessary. Logistics should not be romanticised into a master explanation for every failure. Poor leadership, bad policy, underfunding, corruption, and mis-specified infrastructure all matter. A state can become so enamoured of logistical sophistication that it mistakes movement for strategy and stock for wisdom. The argument here is narrower and stronger. Logistical weakness often converts other weaknesses into visible breakdown. It is the domain in which structural problems become operational facts. That is why ignoring it is so costly. Institutions can survive imperfect forecasting, contested politics, and even some design error longer than they can survive the inability to move what they need when they need it.

A fair counter-example should be acknowledged. Some systems perform remarkably well under strain because logistics has been built into planning from the start. They know which routes matter, which lift assets are scarce, which parts are irreplaceable in the short term, which warehouses need to be regional rather than central, and which maintenance functions cannot be safely outsourced beyond immediate reach. These systems are not magically immune to disruption. They simply respect movement as a governing variable rather than as a technical afterthought. That respect changes how plans are written, how contracts are structured, and how readiness is measured. It also changes public experience because promises are more likely to survive contact with damaged geography.

This returns us to the idea of logistics as the hidden determinant. What is hidden is not the trucks, ports, depots, or repair crews themselves. What remains hidden is their constitutional role inside practical governance. They decide whether authority can travel. They decide whether funding can become function. They decide whether resilience remains a speech or becomes a service. Once this is understood, many debates over readiness look different. The relevant question is no longer merely whether a system has a plan, a budget, or a designated critical asset. It is whether the route to that asset remains open, whether the repair chain is real, whether the spare part exists nearby, whether the crew can arrive, and whether the lift capacity assumed by the plan belongs to the institution or only to its imagination.

The road decided in Puerto Rico, in Texas, and in countless quieter disruptions where the public mostly saw the delay rather than the narrow channel behind it. The unresolved question is not whether institutions appreciate logistics in principle. Most say they do. The sharper question is whether they are willing to treat transport, warehousing, and maintenance as first-order determinants of readiness, and therefore to ask, before the next disruption, which route, depot, contractor, runway, port, bridge, or lifting asset stands between their promise of restoration and the moment the public discovers how little of that promise was ever built to move.

Metrics that do not mislead

In August 2005, when Hurricane Katrina struck the Gulf Coast of the United States, one of the most painful discoveries was that formal preparedness and actual performance were not the same thing. The place was the Gulf Coast. The date is easy to verify. Plans existed, responsibilities existed, compliance language existed, and still the visible strain of evacuation, sheltering, logistics, and public communication exposed how little some official indicators had captured about the system's true readiness under load. The dashboard looked calm.

That image matters because institutions like metrics for reasons that are perfectly understandable. Metrics promise comparability, oversight, funding discipline, and the appearance of control. They help superiors ask whether subordinates are performing, treasuries decide what deserves money, auditors decide what deserves scrutiny, and ministers decide what can be announced with confidence. Yet the same attraction can become dangerous. A metric can make a system legible while also making it easier to misread. When the chosen indicator rewards completion of plans, frequency of exercises, or nominal compliance with reporting rules more than the capacity to function during compound strain, an institution may become increasingly well measured and increasingly poorly prepared at the same time. The dashboard looked calm again, and the calm was part of the problem.

This is the central argument. Measurement is governance, not just description, because what a system chooses to count shapes what it notices, what it funds, what it explains upward, and what it quietly neglects. The concept to name here is performative measurement. A metric does not simply record organisational reality; it helps create the organisational incentives that determine which realities are preserved and which are allowed to thin. The implication is exacting: superficial indicators can reward paperwork over performance, and once they are embedded in audit culture, budget routines, and political reporting, they can make institutions more confident in precisely the areas where they should feel least secure.

This is why paperwork metrics remain so attractive even when their limits are widely understood. They are easy to collect, easy to compare, and easy to defend in public. A ministry can say that every critical facility has an approved continuity plan, every agency has completed a training cycle, every region has filed its resilience assessment, every operator has passed a reporting requirement. None of these claims is meaningless. The problem lies in what they omit. A continuity plan may assume transport access that fails under simultaneous regional disruption. A training cycle may rehearse one hazard at a time while real stress arrives through overlap. A resilience assessment may record assets without testing interdependencies. Reporting compliance may reveal administrative discipline while concealing operational brittleness. The institution can then look orderly in the spreadsheet while remaining fragile in the field.

A useful example comes from hospital and health system preparedness, where indicators often privilege what is easy to audit rather than what is hardest to sustain. It is simpler to record that a hospital possesses a backup power plan than to ask whether fuel delivery can continue during regional disruption, whether communications systems remain functional when the primary network fails, whether staff can reach the site under transport stress, or whether digital records and referral systems continue to operate when several infrastructure layers degrade together. That is why communications infrastructure can become critical to healthcare continuity in ways older metrics may not capture. A hospital may meet formal preparedness standards and still fail operationally because the measures did not follow the real dependency chain. The issue is not that planning was absent. It is that measurement pointed attention in the wrong place.

A second example lies in utility resilience. Utilities often report restoration times, asset hardening progress, inspection completion, and formal emergency readiness. These indicators matter, but they can also flatter performance if they are divorced from compound conditions. A restoration metric drawn from isolated storms may hide what happens when road access, warehousing, repair crews, and fuel logistics are all under stress at once. An asset-hardening metric may ignore that substations, communications links, and key repair routes depend on one another in ways the measure does not reflect. An inspection-completion target may reward routine process while failing to reveal whether the system can still restore critical services when the parts, crews, or lift capacity needed for repair are regionally scarce. Again, the problem is not measurement itself. It is measurement that mistakes what is administratively countable for what is strategically decisive.

This is why useful metrics need sensitivity to compound events. A serious readiness indicator should not only ask whether a component survives under ordinary strain. It should ask what the measure assumes about surrounding conditions. If a facility's continuity score presumes functioning roads, available staff, intact communications, and timely resupply, then those assumptions should be visible rather than buried. If an emergency response target reflects performance under single-hazard events only, then the metric should not be allowed to masquerade as a general statement of resilience. Compound-event sensitivity does not require fantastical modelling. It requires honesty about the fact that systems fail through interaction as often as they fail through single-point weakness.

Clarity on assumptions is therefore not a technical nicety. It is part of whether a metric deserves power. Every measure rests on assumptions: about what counts as acceptable downtime, about which dependencies remain outside scope, about whether neighbouring systems will still function, about what sort of public behaviour is expected, about how much staff absence can be tolerated, about which supply routes remain open. A metric that hides these assumptions encourages dangerous overconfidence because it presents contingent performance as though it were intrinsic strength. A more honest metric does not merely state a score. It reveals the conditions under which the score holds and, by implication, the conditions under which it may cease to mean what decision-makers want it to mean.

This is also why linkage to decision rights matters so much. A measure that cannot trigger review, revision, expenditure, escalation, or public explanation is often little more than institutional decoration. It may satisfy reporting requirements. It may still fail as governance. Useful metrics change behaviour because someone knows what will happen if the indicator worsens, if it crosses a threshold, or if two indicators deteriorate together. Without that link, institutions can accumulate exquisite monitoring systems and still drift. They know more, but nothing compels movement. Measurement then becomes a substitute for action rather than a bridge to it. The metric is admired, discussed, and filed, while the vulnerability it tracks remains politically easier to tolerate than the action its trend would justify.

Still, caution is necessary. It would be a mistake to imagine that more complex metrics are automatically better. Complexity can hide as much as simplicity if it produces models that few people understand and fewer still trust enough to act upon. Institutions can become infatuated with dashboards dense enough to impress auditors while leaving front-line managers unsure which signal truly matters. A useful measure should therefore be rich enough to reflect interdependence and compound strain, yet clear enough that decision-makers can explain what it shows, what it assumes, and what they are empowered to do when it changes. The answer to misleading simplicity is not decorative complexity. It is disciplined relevance.

A fair counter-example should also be admitted. Some seemingly superficial indicators do useful work because they keep minimum standards from collapsing. Documentation, training frequency, asset registers, and inspection completion all provide necessary foundations. A system without them is usually worse, not better. The stronger argument is that foundations should not be mistaken for readiness in full. A hospital that lacks a continuity plan is plainly exposed. A hospital with a continuity plan may still be exposed if the plan's real dependencies and triggers have not been tested or measured in ways that can influence decision. Useful governance therefore layers metrics. It does not discard basic compliance measures, but neither does it allow them to monopolise attention.

The political problem is that misleading metrics often survive because they serve several masters at once. They give auditors a clean trail, ministers a clear claim, agencies a manageable reporting burden, and treasuries a basis for comparing programmes. Measures tied to actual performance under compound stress are harder. They expose assumptions, reveal uncomfortable dependencies, and often imply higher near-term cost. An honest metric may tell leaders that a celebrated programme is weaker than its public narrative suggests, or that a cheap efficiency gain has quietly increased restoration time under regional disruption. Institutions do not rush towards such truths. They have to be pushed, or designed, into measuring what may later embarrass them.

This is why the section's larger claim matters so much. Measurement is governance because it structures attention before crisis and allocates seriousness before breakdown. If the metric tracks the wrong thing, funding will drift towards visible reassurance rather than practical capability. If the metric ignores compound strain, planning will continue to flatter systems that survive isolated disturbance and fail under overlap. If the metric remains disconnected from decision rights, monitoring will produce awareness without accountability. Once that is understood, debates about readiness indicators cease to be technical housekeeping. They become debates about what kind of reality the state is willing to confront before the public does it for them.

The dashboard looked calm before Katrina, and many dashboards have looked calm before quieter failures that never reached international headlines but still exposed the same institutional weakness. The unresolved question is not whether organisations can count more things. They can. It is whether they are willing to count in ways that make their hidden assumptions visible, their interdependencies harder to ignore, and their funding choices less flattering - because only then does a metric stop being a comfort and start becoming a tool for reform.

Chapter 14

Strategic Prioritisation under Feedback Risk

How does a government choose what to protect first when nearly every vital system now depends on several others to keep working? The question sounds managerial until the consequences become visible. Protect the grid and fuel may still fail. Protect transport and communications may still collapse. Protect hospitals and supply routes may still leave them unable to function. The old comfort of ranking assets one by one begins to break down when failure moves through links rather than through isolated objects.

That is why the language of prioritisation so often provokes frustration. Citizens hear it as abandonment. Officials fear it will sound like confession. Politicians prefer the kinder fiction that everything important can be defended at once if the system is merely better led. Yet under feedback risk, that fiction becomes harder to sustain. Interdependence does not abolish the need to choose; it makes the consequences of choosing less visible at first and more decisive later. The queue is already forming.

The deeper difficulty is moral as much as technical. Prioritisation under strain is never just a question of engineering sequence or budget arithmetic. It decides whose disruption is treated as tolerable, which failures are allowed to spread while another node is stabilised, and how candidly institutions are willing to describe the trade-offs built into their own plans. A system that refuses to state those trade-offs does not become fairer. It becomes harder to scrutinise and easier to disappoint. The queue is already forming, whether or not anyone is willing to name it.

This chapter turns on that tension between connectedness and choice. Earlier arguments in the book have shown that methane feedback risk matters not simply because it may intensify physical disruption, but because it shortens decision windows, strains logistics, rearranges criticality, tests institutional learning, and exposes the limits of deterrence language in a world governed increasingly by cascading interdependence.

The problem now is sharper. Once those limits are recognised, what sort of prioritisation remains legitimate, and what sort of institution can carry it without mistaking improvisation for strategy?

The stakes are high because societies rarely fail only at the moment of impact. They fail earlier, in the quiet distribution of attention. What leaders choose to measure, rehearse, protect, and explain in advance determines whether disruption remains bounded or begins to travel. A state that is unclear about priority will often discover that its resources have been spread thinly across symbolic commitments while the decisive dependency was left underprotected. A state that is brutally clear without public legitimacy may preserve one chain of continuity while destroying trust in the wider order. Neither outcome is stable. The issue is not simply efficiency. It is whether the act of choosing can be made intelligible enough that even unwelcome decisions retain some claim to reason.

That problem also reaches beyond national borders. Climate-linked stress does not wait for strategic consensus, and practical safety cannot depend on perfect political harmony. Some forms of coordination remain necessary even where interests diverge, yet they only work when the participating institutions know exactly what function they are trying to preserve. Too much unity is often pretended. Too little cooperation is often tolerated until harm proves more expensive than coordination would have been. The same discipline that governs domestic prioritisation - clarity about dependencies, consequences, and authority - also governs whether states can work together without indulging fantasy about shared purpose.

The movement here follows that discipline from hard choice to institutional form and then outward to the international arena before returning to the book's central claim. It begins where every real planning system eventually arrives: not at the question of whether everything matters, but at the harder question of what must be stabilised because so much else depends on it. From there attention shifts to the kinds of institutions that can function under uncertainty without requiring false certainty from the people inside them. The argument then crosses borders to ask what limited coordination can still achieve when deeper strategic agreement is absent, before returning to methane itself as the variable that forces preparedness to be understood as a matter of stability rather than environmental housekeeping.

What joins these concerns is a single demand for candour. Not optimism. Not fatalism. Candour about dependency chains, about the limits of universal protection, about the difference between cooperation and consensus, and about the way methane feedback risk alters the planning value of time. Once methane is treated as a strategic variable rather than a side issue in climate rhetoric, budgets, infrastructure design, logistics, communication, and institutional revision stop looking like separate policy fields. They become parts of one larger question: how much disorder can a state absorb before service failure, fiscal strain, and public mistrust begin to reinforce one another faster than governance can recover.

That is why the book ends here rather than in a catalogue of technical fixes. The final problem is not lack of warning. It is whether institutions can still decide responsibly when certainty is incomplete, interdependence is thick, and the political temptation to promise everything remains stronger than the administrative ability to deliver it. The sharper question, then, is not whether prioritisation can be avoided. It is whether the priorities already embedded in the system are honest enough to be defended once the next chain of failures reveals what the state had, in practice, chosen all along.

Prioritisation when everything is connected

In August 2005, when Hurricane Katrina struck the Gulf Coast of the United States, one of the hardest lessons was not simply that many systems failed at once, but that no government could protect everything at once even if it wanted to. The place was the Gulf Coast. The date is easy to verify. Roads, fuel, shelter, policing, hospitals, communications, water systems, and public order all competed for attention under conditions in which delay in one domain quickly worsened failure in another. The queue was already there.

That is the right place to begin because modern planning still clings to a comforting fiction: if enough risks are named and enough assets are labelled important, responsible government can avoid choosing among them until a crisis forces the matter. The fiction is politically useful. It spares leaders the pain of admitting that interdependence does not abolish prioritisation but makes it harsher. It also produces one of the most common failures in resilience planning, which is to treat all risks as equally urgent in principle and therefore to remain unclear about what must actually be protected first when several critical functions are threatened together. The queue was already there again, not because officials were indifferent, but because the system had not decided what kind of order it intended to defend first.

This is the central tension. When everything is connected, prioritisation becomes more necessary and more controversial at the same time. The concept to name here is consequential priority. It means ranking action not by symbolic importance, political noise, or the abstract dignity of a sector, but by the consequences that follow if a specific dependency fails. The implication is exacting: a government that cannot explain why one node, route, stock, or service receives protection before another is not avoiding politics. It is allowing politics to occur in a less honest form, through improvisation, confusion, and later claims that no real choice had ever been made. Under feedback risk, that kind of vagueness becomes increasingly expensive.

This matters because interdependence changes what urgency means. In older planning habits, urgency could be attached more directly to visible sectors. Hospitals are urgent. Power is urgent. Water is urgent. Communications are urgent. Transport is urgent. All true, and still insufficient. Under compound strain, urgency belongs increasingly to what holds several of these functions together. A fuel distribution network, for example, may look like one infrastructure among many until one notices how quickly health services, food logistics, backup generators, repair crews, emergency vehicles, and public order all begin to depend upon it once the ordinary timetable fails. Protecting fuel then becomes less a sectoral preference than a way of protecting several services at once. That is what good prioritisation sees and bad prioritisation obscures.

The common planning failure lies in refusing to distinguish between moral importance and strategic sequence. A shelter, a clinic, a substation, a warehouse, and a communications hub may all be morally important. They do not all occupy the same place in a failure chain. If a communications hub fails, healthcare continuity may degrade because records, referrals, remote consultation, ambulance dispatch, and household contact all begin to fray together. If fuel fails, the apparent autonomy of several sectors can vanish quickly. If a key transport route fails, restoration plans funded and authorised at high level may remain physically inert. Prioritisation becomes clearer when institutions ask not "which asset matters most in the abstract?" but "which point of protection or restoration keeps the greatest number of other functions from sliding into difficulty?"

That is why dependency chains are so useful. They force a system to move beyond static lists of importance and towards dynamic sequences of consequence. A dependency chain reveals what relies on what, under what conditions, with what lag, and with what substitutes still realistically available. Such analysis rarely flatters the existing organisation chart. It often shows that sectors treated as separate in budgeting or regulation are functionally inseparable under stress. It may show that a modest depot matters more than a prestigious building, or that a secondary road matters more than a formally critical site because the road is what allows the critical site to be supplied, staffed, and repaired. This is not romantic systems language. It is a disciplined way of making trade-offs visible before the crisis makes them brutally obvious.

A brief example makes the logic concrete. Suppose a regional government faces overlapping disruption: power instability, supply chain strain, and mounting public anxiety. It has limited crews, limited transport, limited political attention, and no possibility of fully restoring all systems at once. One option is to spread effort thinly across many sectors in the name of fairness. Another is to focus first on keeping fuel distribution functioning: securing storage sites, protecting delivery routes, prioritising repair support for depots and loading systems, and ensuring emergency allocation rules are operational. That choice may look narrow or unpopular, especially to communities whose immediate hardship lies elsewhere. Yet if fuel continuity keeps ambulances moving, generators running, repair crews deployed, food distribution viable, and water treatment functioning, the choice may preserve more practical order than a more egalitarian but less strategic distribution of effort. The hard part is not making such a decision privately. It is defending it publicly without insulting those who bear the delayed benefit rather than the immediate relief.

This is where legitimacy enters. Clear trade-offs are politically dangerous, but obscured trade-offs are often worse. When authorities refuse to explain why one chain of support is being stabilised before another, citizens do not experience the choice as neutral. They experience it as arbitrariness, favouritism, or neglect. Clarity does not remove anger. It can, however, make anger more governable because people can at least see the reasoning. A system that says openly, in effect, that protecting this route, depot, or node is meant to preserve several downstream services is taking a risk with candour. That risk is often worth taking. Legitimacy grows not from universal agreement but from the visible presence of reasons that connect sacrifice to continuity in a way the public can inspect and contest.

Still, caution is necessary. Dependency chains can be misused. Some institutions will overstate the centrality of their own sector, presenting every inconvenience as a national hinge. Others will become so enamoured of cascading logic that they neglect equity, assuming that the most technically connected asset must always outrank the most socially fragile community. That would be a serious mistake. Prioritisation is not morally exhausted by systems analysis. It still requires judgment about who is most exposed, who can wait, who has alternatives, and which harms become politically or ethically intolerable if deferred. The stronger claim is narrower. Dependency analysis does not decide everything. It helps prevent decisions from being made in ignorance of their wider consequences.

A fair counter-example should also be admitted. In some emergencies, broad distribution of assistance is more stabilising than narrow concentration because public order, trust, or survival depends on visible fairness. A strategy that is technically elegant but publicly illegible can backfire if people conclude that the state is protecting systems while abandoning persons. This matters because prioritisation always occurs in a social field, not a diagram. A government that protects fuel distribution brilliantly while allowing communication about the choice to fail may preserve some operational continuity and still suffer severe legitimacy loss. The answer is not to abandon prioritisation. It is to accept that good prioritisation has to be both technically coherent and publicly explainable.

This is why the reader should treat prioritisation as a question of design as much as of judgment. Has the institution mapped failure consequences well enough to know which nodes matter most under compound strain? Has it identified where one protected function preserves several others? Has it decided in advance which choices require political explanation rather than technical concealment? Has it built a language for saying, without euphemism, that some protections are being sequenced because the alternative would expose more people to faster and wider failure? These are the questions that separate serious strategic prioritisation from the gentler fiction that every important thing can be defended at once if only the plan is comprehensive enough.

The queue was already there in 2005, and it appears in every complex crisis whether officials admit it or not. The unresolved question is not whether governments will have to choose among connected risks. They will. It is whether those choices will be made through a transparent understanding of dependency and consequence, or through the more damaging habit of pretending that all priorities remain equal until the moment the public learns, too late and by experience alone, which ones were real.

Designing institutions for uncertainty

In October 1962, during the Cuban Missile Crisis, the most consequential failures would not have come only from bad intentions or missing information. They would also have come from ambiguity about who could decide, how fast information moved, and whether revised judgment could enter the system before events outran it. The date is easy to verify. The places are obvious. The enduring lesson is administrative as much as strategic: uncertainty punishes institutions that require clarity before they can act, and it rewards those able to move responsibly while the picture is still incomplete. The hinge is small.

That image matters because public debate often treats uncertainty as though it were mainly a problem of knowledge. Gather better data, improve the model, sharpen the warning, and governance will follow. Sometimes it does. More often, uncertainty reveals something harsher. Institutions may know quite a lot and still perform badly because mandates overlap, plans are too rigid, updating looks like weakness, and learning remains trapped in reports rather than converted into revised authority. The hinge is small again, because what decides performance is often not the quantity of information in the system but the design of the system that must turn information into judgment.

This is the central argument. Institutions that perform under uncertainty share certain properties, but they do not add up to a universal blueprint. The concept to name here is adaptive fit. It refers to the degree to which a governance structure matches the tempo, ambiguity, and interdependence of the risks it is meant to manage. The implication is exacting: one should judge institutions less by whether they appear comprehensive and more by whether they can revise, coordinate, and decide without waiting for the world to become tidier than it ever will be. Under feedback risk, the question is not whether an institution has a plan. It is whether the institution is built to remain governable once the plan meets contradiction.

Clear roles are the first property worth examining, not because hierarchy is always desirable, but because uncertainty amplifies the cost of ambiguity. In ordinary periods, overlapping authority can look flexible. In crisis, it often becomes an invitation to delay. If several offices can warn, fund, approve, or defer, each may wait for the others to shoulder the political risk of moving first. A structure fitted to uncertainty does not eliminate overlap entirely. It makes visible who owns what decision at what stage, who can escalate, who can override, and who must explain a refusal to act. This is not bureaucratic neatness for its own sake. It is a way of reducing the time lost to institutional hesitation while preserving accountability for difficult calls.

One can see the importance of role clarity in public health and emergency management, where confusion over scientific advice, operational instruction, and political reassurance has repeatedly weakened otherwise capable systems. A technical body may know when a threshold has become meaningful. A political executive may hesitate because the public case feels fragile. A regional authority may face the practical burden without possessing the legal discretion to adjust the response. None of these actors is necessarily incompetent. The structure may simply be mismatched to the risk landscape. A diagnostic question therefore becomes unavoidable: when conditions worsen, does the system reveal who decides, or does it reveal that several institutions had assumed someone else would own the hardest part of the choice?

Modular plans form the second property. A plan built for certainty usually assumes that the whole architecture will operate roughly as designed. A plan fitted to uncertainty assumes that parts may fail, conditions may diverge across regions, and revision may be necessary before full implementation. Modularity allows institutions to substitute, suspend, or re-sequence components without treating every change as systemic defeat. This matters because feedback risks rarely arrive in one neat package. Logistics may strain before communications fail. Fuel may remain available while staffing collapses. One region may face a slow grind while another faces abrupt overlap. A modular institution does not require the entire machine to remain intact before any part can work. It protects continuity by refusing the fantasy that the full original script will survive first contact.

The value of modularity is clearest in systems that must coordinate across sectors. A rigid continuity plan may assume stable transport, stable communications, and ordinary procurement all at once. A modular one can continue operating if one support function degrades, because it has already defined fallback pathways, temporary substitutions, and reduced but credible service modes. This is why modularity should not be confused with fragmentation. Fragmentation disperses authority without coherence. Modularity preserves coherence while allowing partial adjustment. The former multiplies confusion. The latter preserves room to move. A government facing uncertainty should not ask only whether plans exist, but whether they can survive partial failure without forcing the entire system into all-or-nothing politics.

Credible updating is the third property, and perhaps the most politically difficult. Institutions often understand that conditions change. What they fear is the appearance of inconsistency. A government that revises guidance, reallocates resources, or changes strategic emphasis may worry that it looks weak, indecisive, or deceptive. Yet under uncertainty, the refusal to update is usually more damaging than the update itself. What matters is whether revision has been normalised as part of the governing method rather than treated as embarrassment. An institution fitted to uncertainty can say, in effect, that this was the best available judgment under the previous conditions, those conditions have altered, and therefore the next step changes as well. That kind of candour strengthens trust when the system has already built a record of explaining how revision works.

Still, caution is necessary. Updating can become a theatre of flexibility if institutions treat every new piece of information as a reason to reset direction. That is not adaptability. It is loss of nerve. The issue is not constant movement. It is disciplined revision anchored to clear triggers, intelligible reasons, and a stable sense of purpose. A good diagnostic framework therefore asks not whether an institution changes often, but whether it can change *coherently*: whether staff know what evidence matters, whether the public can see why guidance shifted, whether previous plans can be revised without bureaucratic humiliation, and whether the authority to update is lodged where the relevant knowledge actually appears.

Learning systems complete the picture because uncertainty punishes institutions that mistake reporting for learning. Many organisations produce after-action reviews, inquiry reports, or lessons-identified documents. Fewer convert them into revised mandates, altered procurement rules, different staffing assumptions, or new thresholds for escalation. The distinction is decisive. A learning system is not one that remembers everything. It is one that changes something material when the same weakness appears more than once. That may be a legal amendment, a budget line, a revised role for a regional office, a different stock pattern, or a new trigger linked to decision rights. If the lesson remains admirable prose, the institution has documented its vulnerability rather than reduced it.

A fair counter-example deserves notice. Some relatively centralised systems perform under uncertainty with less modularity and less formal updating because they possess deep professional competence, stable administrative culture, and strong informal coordination. These cases matter because they prevent the argument from hardening into one organisational doctrine. Yet even such systems usually rely on tacit knowledge, trusted intermediaries, and unwritten habits that can degrade under turnover, polarisation, or repeated strain. The point is not that every institution must look the same. It is that every institution should be judged against the same harder question: does its actual design match the uncertainty it faces, or does it still reflect a calmer world in which risks arrived one at a time and authority could afford to move slowly?

This is why the section avoids blueprint claims. A blueprint flatters the idea that a single model can be transplanted from one political and administrative setting to another. A diagnostic framework asks something more mature. Where does ambiguity accumulate? Which functions depend on tacit coordination rather than authorised pathways? How reversible are key decisions? Can parts of the system continue under degraded conditions? Does learning alter anything beyond language? Can public updating occur without reputational collapse? These questions do not produce one perfect institutional design. They do something better. They expose whether the institution's current form is likely to amplify uncertainty or absorb it.

The hinge is small, and that is precisely the problem. Systems often fail not because they lacked grand strategy, but because a modest ambiguity in mandate, a brittle plan structure, or a politically costly revision pathway widened under stress until all the larger ambitions could no longer travel through it. The unresolved question is not whether institutions can be made uncertainty-proof; they cannot. It is whether they are honest enough to examine the points where their own structure still assumes a level of clarity, sequence, and control that the risk landscape has already stopped offering.

International coordination without false unity

In April 1986, after the Chernobyl disaster, radioactive fallout crossed borders faster than diplomatic comfort. The date is easy to verify. The place was the Soviet Union, but the political problem travelled across Europe. States did not need shared strategic aims to discover that one country's emergency had become many countries' monitoring problem at once. They needed data they could compare, warnings they could trust enough to act on, and at least a minimum of practical coordination while suspicion and political embarrassment were still very much alive. The channel stayed open.

That image matters because international coordination is often discussed in one of two equally unhelpful registers. One treats cooperation as a moral ideal that should dissolve disagreement if only leaders are responsible enough. The other treats any meaningful coordination as naive unless broader strategic interests are aligned. Neither view is serious enough for feedback risk. Many hazards do not wait for political harmony. Smoke crosses frontiers. Shipping incidents escalate through shared waters. Supply disruptions spread through connected infrastructure. Monitoring signals become useful only if someone else's system can read them in a compatible way. The channel stayed open again, not because unity had been achieved, but because some functions were too operationally important to be left hostage to perfect agreement.

This is the central argument. States can coordinate on safety, monitoring, and disaster response without pretending that they share a full strategic worldview. The concept to name here is functional coordination. It means cooperation framed around concrete tasks narrow enough to survive disagreement: exchanging data that supports warning, maintaining interoperability for emergency response, preserving channels for technical contact, and clarifying procedures for incidents that would become worse if every state acted alone or guessed at the others' intentions. The implication is exacting: limited coordination can still reduce harm, but only if it is built around specific functions rather than rhetorical declarations of common purpose that political reality cannot sustain.

This matters because false unity can be as damaging as open friction. If institutions describe cooperation in language too grand for the actual relationship, the first geopolitical strain can turn every technical arrangement into a test of loyalty rather than utility. That makes modest but valuable systems more fragile than they need to be. By contrast, coordination designed with a narrower ambition often proves more durable. A common data format for flood alerts, a standing contact point for disaster aviation, compatible communications protocols for search and rescue, or agreed procedures for sharing hazard observations do not require trust in the abstract. They require enough practical confidence that the concrete function is worth preserving even while larger disputes remain unresolved. That is a lower bar. It is also, in many domains, the more honest one.

The practical barriers to such coordination are rarely glamorous. Data standards are one. A warning is only as useful as its legibility on the receiving end. If neighbouring states or agencies measure differently, classify differently, geolocate differently, or package risk information in incompatible systems, then the political promise of coordination collapses into technical noise. Interoperability is another. Emergency response is often discussed as a matter of goodwill, but goodwill does little when radio systems do not align, mapping conventions differ, logistics software cannot exchange information, aircraft procedures conflict, or legal authorities for cross-border assistance remain murky at the moment speed matters most.

Political trust is the third barrier, and in some ways the hardest, because it governs whether technical mechanisms will be used as intended or treated as potential avenues of exposure, blame, or strategic disadvantage.

A useful example lies in Arctic search and rescue, where states with differing strategic interests have still had reason to preserve practical channels because the environment punishes delay more than it rewards posturing. The relevance of Arctic cases is not that they offer a utopia of cooperation. They do not. Their importance lies in demonstrating that narrow operational needs - distress response, hazard monitoring, safe movement, technical communication - can sustain limited coordination even when the wider relationship is burdened by suspicion. The lesson for broader global risk management is precise. Where the physical setting is unforgiving, function-focused cooperation becomes more valuable, not less, because the cost of misperception or procedural incompatibility is so high.

A second example can be found in transboundary meteorological and disaster warning systems more generally. States may disagree profoundly on security, trade, ideology, or regional order and still retain a shared interest in timely warning, comparable observation, and emergency information that neighbours can actually use. The value here is not sentimental. It lies in reducing avoidable harm through compatibility. A storm warning that cannot be parsed across the border, an evacuation advisory that uses incompatible mapping references, or an air quality signal that arrives too late to support hospital or school decisions is not merely a technical inconvenience. It is a governance failure that could have been reduced without requiring deep diplomatic convergence on unrelated matters.

Still, caution is necessary. Limited coordination should not be romanticised as politically effortless. Even narrow technical cooperation can become entangled in wider tensions. States may fear exposing vulnerability through data sharing. They may suspect that monitoring information will be used for reputational attack or strategic leverage. Domestic politics may punish officials seen to cooperate with an adversary, however practical the function. Legal systems may block rapid exchange of information or equipment. Private infrastructure operators may complicate matters further by holding data or assets that governments cannot easily mobilise across borders. A serious analysis has to admit these frictions or else it mistakes aspiration for design.

That limitation strengthens the main point because it forces attention towards function rather than mood. The question is not whether states feel cooperative. The question is whether a specific arrangement can survive ordinary political stress because its purpose is clear, its costs and risks are bounded, and its benefits are visible enough to justify maintenance. A data-sharing channel for hazard monitoring may endure where a broader strategic dialogue fails because the narrower channel is easier to defend domestically and easier to separate from larger disputes. An interoperability protocol for search and rescue may survive because the alternative is so plainly dangerous. Good design begins there, with the cooperation that can be justified in practical terms even by governments that do not trust one another much.

This is why political trust, although essential, should not be treated as a mystical precondition that must be fully present before any useful coordination can begin. In many cases trust is better understood as layered and task-specific. States may lack confidence in one another's intentions on major strategic questions and still trust a particular technical process enough to exchange weather data, share a shipping incident alert, or honour a search-and-rescue contact protocol. That narrower trust is often built through repeated procedural reliability rather than emotional warmth. Messages arrive when promised. Formats remain stable. Contacts respond. Exercises clarify roles. Nothing magical occurs. The practical habit of limited trust thickens, and with it the chance that future coordination will remain possible under pressure.

The connection to earlier discussions of monitoring and institutional design should now be clear. Information only reduces harm when it is understandable, actionable, and linked to decision rights. The same holds across borders. A cross-border warning system that produces incompatible data, unclear thresholds, or no authorised pathway for action is less a cooperative achievement than a shared performance of concern. Likewise, disaster response arrangements that assume goodwill but neglect standards, equipment compatibility, legal triggers, and communication discipline are vulnerable to failure precisely when political strain makes improvisation least safe. International coordination without false unity therefore depends on the same virtues demanded of domestic governance: clarity, procedural honesty, and a refusal to confuse presence of a framework with readiness of a function.

A fair counter-example should also be admitted. Some emergencies do expose the limits of narrow functionalism. There are cases where strategic hostility is so severe that even useful technical coordination erodes, where sanctions, war, espionage concerns, or domestic political pressures narrow contact to the point that no meaningful shared practice can survive. This matters because the argument here is not that practical function can always outlast power politics. It is that where cooperation remains possible, it is often strongest when framed around concrete tasks rather than enlarged into a false story of common purpose. Function does not defeat rivalry. It can, at times, keep rivalry from poisoning every safety-relevant domain equally.

That is the deeper strategic gain. Limited coordination makes the international system less brittle. It does not create friendship. It reduces the odds that misunderstanding, incompatibility, or procedural incoherence will turn a difficult event into a worse one. For feedback risks, that reduction matters greatly because the dangers are often compounding and transboundary. A state protecting its own citizens may still depend on another state's warning discipline, data compatibility, or transport procedure. To admit this is not to weaken sovereignty. It is to recognise that sovereign competence in an interdependent hazard field includes the ability to cooperate selectively without pretending to agree about everything.

The channel stayed open after Chernobyl because, once harm began to travel, function mattered more than diplomatic comfort. The unresolved question is whether states facing the next era of compounding risk can build enough narrow, durable, technically credible coordination that they do not need to wait for false unity before reducing shared danger - or whether they will continue to discover, each time the hazard crosses the border first, how much preventable harm had been stored inside their insistence on treating cooperation as all or nothing.

What changes when methane is priced as security

What changes when a gas usually discussed in inventories and climate negotiations is treated instead as a problem of state continuity? The question is sharper than it first appears. It does not ask whether methane matters morally, scientifically, or diplomatically; those arguments already exist. It asks what happens when methane feedback risk is taken seriously as a planning variable inside the institutions that budget for repair, design infrastructure, organise emergency response, and speak to the public when systems begin to strain. The ledger changes.

That shift matters because security language is often misused to inflate urgency without improving judgment. Here the task is different. To price methane as security is not to declare an enemy where none exists, nor to smuggle environmental concern into a martial frame for rhetorical force. It is to recognise that methane-related feedback risk can alter the pace, clustering, and administrative consequences of disruption in ways that bear directly on territorial reliability, fiscal burden, infrastructure performance, and public trust. The ledger changes again because what was once treated as an externality begins to enter the core accounting of state capacity.

This is the book's central contribution brought to its hardest point. Methane feedback risk matters strategically because it changes the conditions under which ordinary government functions remain governable. The issue is not only warming in the abstract, still less a competition among gases for political attention. The issue is timing, interaction, and consequence. If methane-related processes tighten decision windows, intensify compounding shocks, and raise the likelihood that several systems will strain together, then they belong inside the logic by which governments rank priorities, design assets, hold reserves, rehearse disruption, and explain trade-offs to the public. A planning system that excludes such risk from its strategic calculations is not neutral. It is silently betting on a more forgiving world than the one it may have to govern.

This changes budgets first, though not in the simplistic sense that every ministry must suddenly spend more on everything. The deeper change is in what counts as prudent expenditure. A budget built around stable baselines and isolated emergencies will tend to reward visible output, short-term savings, and repair after failure. A budget that prices methane-related feedback risk as security begins to ask harder questions. Which spending preserves continuity across several systems at once? Which apparently efficient cuts remove buffers that matter most when events overlap? Which maintenance lines, warehousing decisions, staffing margins, or local stocks appear excessive only because the accounting framework still assumes recovery will remain sequential rather than compounding? Once methane enters as a security variable, fiscal judgement becomes less about annual neatness and more about whether the state is buying enough time, slack, and optionality to remain credible when strain thickens.

Infrastructure design changes for similar reasons. The argument is not that every asset must be rebuilt to some maximal standard. That would be neither possible nor wise. The stronger claim is that design decisions begin to look different when the future is no longer treated as a relatively linear extension of the recent past. A road, runway, port, communications system, depot, or health facility is no longer only a capital asset with a service life.

It becomes a commitment about what conditions the state believes it will still be able to manage. To price methane as security is therefore to ask whether infrastructure is being designed for graceful degradation, modular repair, and realistic maintenance under interrupted logistics, or whether it still assumes a tempo of disruption and recovery that methane-related feedback dynamics may render less plausible. The difference is subtle in procurement language and profound in public consequence.

Public communication changes too, and perhaps more than many governments would like to admit. Institutions often treat communication as the final explanatory layer placed over choices made elsewhere. This book has argued otherwise. When risk is compounding, uncertain, and politically charged, communication becomes part of operational capacity. To price methane as security is to accept that credible public explanation of uncertainty, prioritisation, and staged response is itself a preparedness function. Citizens do not need theatrical certainty. They need an intelligible account of why certain systems are being protected first, why some buffers matter even when they appear idle, why updates occur as conditions shift, and why uncertainty is not an alibi for drift. A state that cannot speak about these matters without alternating between alarm and evasion will struggle to preserve trust at precisely the moment trust becomes a logistical resource.

This is where the structured way of thinking the book has been building comes into focus. Stability under compounding shocks cannot be understood by looking at hazards alone. It has to be read through four linked questions. What dependencies matter most when systems are under load? Which institutional arrangements can still decide and revise under uncertainty? Where do logistics, buffers, and local capacities determine whether plans remain real? And which forms of limited coordination, domestic and international, reduce harm even when deeper consensus is absent? Methane feedback risk does not replace these questions. It sharpens them. It acts as a forcing variable that reveals whether a state's apparent preparedness rests on robust design or on assumptions about timing and separability that may no longer hold.

A clear example lies in fuel distribution, which earlier discussion used to illustrate dependency-based prioritisation. If methane-related feedback risk is treated as strategic, fuel is no longer simply a commodity in a market or a utility within an energy portfolio. It becomes a continuity node whose protection can preserve emergency vehicles, backup power, food logistics, communications maintenance, and repair operations across several sectors at once. The point is not that fuel always comes first. It is that pricing methane as security teaches institutions to look for such leverage points: the places where one defended function protects many others. This changes how budgets justify stocks, how infrastructure is protected, how restoration is sequenced, and how difficult choices are explained before the public learns about them through breakdown.

Still, caution is necessary. To price methane as security is not to claim certainty about every feedback pathway or to pretend that institutions can forecast the exact pace and sequence of future disruption. Much remains uncertain. Threshold behaviour, regional variation, interaction with other climate processes, and the political consequences of physical strain all resist neat prediction. A serious account must preserve that uncertainty. The argument here is not for deterministic planning. It is for decision under uncertainty that is honest about what cannot be settled and disciplined about what can. Governments do not need precise foresight to know that buffers, modularity, clear decision rights, dependency mapping, and credible updating perform better than brittle efficiency and vague optimism when the future may tighten rather than relax.

A fair counter-example also deserves mention. Some governments and institutions will argue, not unreasonably, that they already plan for severe weather, infrastructure stress, and civil contingency without explicitly framing methane as a security variable. Sometimes that planning is substantial and useful. Yet the framing still matters because names organise attention. Where methane remains categorised mainly as a climate reporting issue, it competes for space in environmental silos. Where it is treated as part of strategic planning, it begins to alter how treasuries read maintenance, how infrastructure authorities read design life, how civil contingency systems read concurrency, and how public communication is prepared for uncertainty that is not temporary noise but a persistent feature of the operating environment. The change is therefore not only semantic. It is institutional.

This is why the book ends not with a programme but with a standard of judgment. What remains uncertain should not be hidden. What can still be responsibly decided should not be deferred behind uncertainty it does not require. Governments can decide whether their priority systems are dependency-based or politically improvised. They can decide whether their institutions are built to revise under stress or to defend old assumptions until failure becomes undeniable. They can decide whether metrics point towards real continuity or towards paperwork that flatters performance.

They can decide whether communication treats citizens as partners in difficult trade-offs or as audiences to be reassured until reassurance breaks. None of these decisions requires a final answer to every scientific question about methane feedbacks. They require seriousness about what strategic preparation means when the future may become less forgiving faster than the administrative state prefers.

The ledger changes when methane is priced as security, but that is not the final change. The deeper change is in what counts as responsible government. No state can abolish uncertainty, prevent every compound shock, or guarantee continuous stability across every domain. The unresolved question is whether it can learn to govern with enough candour, enough discipline, and enough structural realism that when the next period of strain arrives, it is not merely measuring instability more precisely than before, but has already decided which forms of continuity it will truly defend and what kind of honesty that defence will demand.

About the Author

Imani Z. Okere is a nonfiction writer working at the boundary where climate risk meets the practical limits of governance. Her work is driven by a simple conviction: large systems do not fail only because hazards intensify, but because institutions inherit assumptions about time, stability, and control that no longer hold. With an editorial voice shaped by academic nonfiction, she is interested in how states translate knowledge into action when the knowledge is probabilistic, politically contested, and morally charged.

Okere's approach favours clear concepts over slogans. She is drawn to the places where policy language becomes evasive: the gap between "resilience" as a public promise and resilience as staffing levels, maintenance schedules, procurement rules, and credible communication. She writes with respect for scientific uncertainty while insisting that uncertainty is not the same as insignificance. Across her work, a recurring thread is the geopolitics of the high latitudes, where infrastructure meets fragile ground and strategic narratives meet daily service provision.

Rather than treating security as a narrow military category, Okere treats it as a question of state capacity under stress: the ability to keep people safe, keep systems functioning, and keep decisions legitimate when shocks compound. Her guiding sensibility is comparative and historically minded, attentive to the way past planning doctrines still shape present-day institutions. She writes for readers who want rigorous analysis without technocratic fog: students, analysts, and general audiences seeking to understand how climate feedback risks change what responsible governance requires.

www.ingramcontent.com/pod-product-compliance
Lightning Source LLC
Chambersburg PA
CBHW021216250926
51852CB00006B/35
* 9 7 8 9 3 7 7 9 4 0 0 8 9 *